ETHICS AT WAR

This book debates competing approaches to ethical decision-making for members of the armed forces of liberal democratic states.

In this volume, four prominent thinkers propose and debate competing approaches to ethical decision-making for military personnel. Deane-Peter Baker presents and expounds the 'Ethical Triangulation' model, an ethical decision-making method he has employed through much of his career as an applied military ethicist. Rufus Black advocates for a natural law-based approach, one which has heavily influenced the framework formally adopted by the Australian Defence Force. Roger Herbert outlines the 'Moral Deliberation Roadmap', the moral reasoning framework recently adopted by the US Naval Academy. Iain King then sets out a model of quasi-utilitarian decision-making developed in several post-conflict settings and refined at the UK's Royal College of Defence Studies. After the opening chapters in which each author outlines their favoured decision-making approach, the four contributors then evaluate each other's proposals, often critically. Philosopher David Whetham offers some concluding thoughts in which he summarizes areas of agreement between the authors, identifies key areas of difference, and suggests directions for future research.

This book will be of great interest to students of military ethics, the ethics of war, moral philosophy, and International Relations, as well as military professionals.

Deane-Peter Baker is Associate Professor in the School of Humanities and Social Sciences at the University of New South Wales, Canberra. He is Director of the UNSW Military Ethics Research Lab and Innovation Network (MERLIN) and a co-convenor (with Professor David Kilcullen) of the UNSW Canberra Future Operations Research Group.

Rufus Black is Vice Chancellor and President of the University of Tasmania. He has held a range of senior executive, academic, board and advisory roles in the public, private, and education sectors in Australia including conducting major reviews for the Australian government on Defence and Security matters.

Roger Herbert, following a 26-year career as US Naval Special Warfare Officer, joined the faculty of the US Naval Academy, where he served until 2021 as the Robert T. Herres Distinguished Military Professor of Ethics and Director of the USNA's core ethics course.

Iain King is an author and defence expert with an extensive background in both ethics and conflict work; most recently he worked as Director of NATO's Mission in Iraq. He led the UK's government research programme into conflict and was a Fellow at the US-based think tank, the Center for Strategic and International Studies (CSIS), as well as being an in-studio commentator for CNN and BBC.

War, Conflict and Ethics

Series Editors:
Michael L. Gross
University of Haifa
and
James Pattison
University of Manchester

Ethical judgments are relevant to all phases of protracted violent conflict and inter-state war. Before, during, and after the tumult, martial forces are guided, in part, by their sense of morality for assessing whether an action is (morally) right or wrong, an event has good and/or bad consequences, and an individual (or group) is inherently virtuous or evil. This new book series focuses on the morality of decisions by military and political leaders to engage in violence and the normative underpinnings of military strategy and tactics in the prosecution of the war.

Distributing the Harm of Just Wars
In Defence of an Egalitarian Baseline
Sara Van Goozen

Moral Injury and Soldiers in Conflict
Political Practices and Public Perceptions
Tine Molendijk

The Empathetic Soldier
Kevin R. Cutright

Law, Ethics and Emerging Military Technologies
Confronting Disruptive Innovation
George Lucas

Ethics at War
How Should Military Personnel Make Ethical Decisions?
Deane-Peter Baker, Rufus Black, Roger Herbert and Iain King

For more information about this series, please visit: www.routledge.com/War-Conflict-and-Ethics/book-series/WCE

ETHICS AT WAR

How Should Military Personnel Make Ethical Decisions?

Deane-Peter Baker, Rufus Black,
Roger Herbert and Iain King

Routledge
Taylor & Francis Group

LONDON AND NEW YORK

Cover image: Getty Images © zabelin

First published 2024
by Routledge
4 Park Square, Milton Park, Abingdon, Oxon OX14 4RN

and by Routledge
605 Third Avenue, New York, NY 10158

Routledge is an imprint of the Taylor & Francis Group, an informa business

British Library Cataloguing-in-Publication Data
A catalogue record for this book is available from the British Library

Library of Congress Cataloging-in-Publication Data
Names: Baker, Deane-Peter, author. | Black, Rufus, author. | Herbert, Roger G.,
    Jr., author. | King, Iain, 1971– author.
Title: Ethics at war : how should military personnel make ethical decisions? /
    Deane-Peter Baker, Rufus Black, Roger Herbert and Iain King.
Other titles: How should military personnel make ethical decisions?
Description: Abingdon, Oxon ; New York, NY : Routledge, 2024. | Series:
    War, conflict and ethics | Includes bibliographical references and index.
Identifiers: LCCN 2023038183 (print) | LCCN 2023038184 (ebook) | ISBN
    9781032321219 (hardback) | ISBN 9781032321202 (paperback) | ISBN
    9781003312925 (ebook)
Subjects: LCSH: Military ethics—Western countries. | Command of troops. |
    Decision making—Moral and ethical aspects. | War—Decision-making. |
    War—Moral and ethical aspects.
Classification: LCC U22 .B268 2024 (print) | LCC U22 (ebook) |
    DDC 172/.42—dc23/eng/20231020
LC record available at https://lccn.loc.gov/2023038183
LC ebook record available at https://lccn.loc.gov/2023038184

ISBN: 978-1-032-32121-9 (hbk)
ISBN: 978-1-032-32120-2 (pbk)
ISBN: 978-1-003-31292-5 (ebk)

DOI: 10.4324/9781003312925

Typeset in Times New Roman
by Apex CoVantage, LLC

To the men and women of the Australian, United Kingdom, and United States Armed Forces, and to the leaders, mentors, and teachers who guide them on their paths of honourable service.

CONTENTS

ACKNOWLEDGEMENTS

Deane-Peter Baker – My opening chapter draws on earlier discussions of Ethical Triangulation in Deane-Peter Baker (ed) *Key Concepts in Military Ethics* (UNSW Press 2015, pp. 33–38) and Deane-Peter Baker *Morality and Ethics at War: Bridging the Gaps between the Soldier and the State* (Bloomsbury Academic 2020, pp. 143–148). I would like to acknowledge the critical role his former colleagues at the US Naval Academy played in shaping his thinking about military ethical decision-making, particularly Captain Rick Rubel (USN, Ret.).

Rufus Black – I would like to thank the two other panel members of the Afghanistan Inquiry Implementation Oversight Panel, Vivienne Thom and Rob Cornell, and the head of its Secretariat Sharon Dean for their engagement and support during important discussions we had within Defence about Australian Defence Force's (ADF) approach to ethics. I would also like to thank those in the ADF for all the discussions we had and for their profound commitment to developing a deeply considered and practical form of military ethics. Finally, thank you to my office team who enabled me to do this work while being a busy Vice Chancellor and to Kieren Rix on my team for his feedback and proofing.

Roger Herbert – Credit for designing the deliberative framework I describe in Chapter 3 belongs primarily to three US Naval Academy philosophers: Chris Eberle, Marcus Hedahl, and Michael Skerker. Also deserving credit are five exceptional naval officers whose combined decades of military experience focused the curriculum and ensured its practical utility: Michael Good, Danielle Litchford, Maria 'MJ' Pallotta, Doug Rau, and Tom Robertson. Additionally, credit belongs to my USNA Department Chair, Joe McInerney, who directed this effort. Finally, I'm grateful to those who lent a critical eye to the chapters I authored. My thanks to John Buford, Bruce Dalcher, Bob Gusentine, Roger Herbert (Sr.), David Nartker, Bob Schoultz, and Sandy Stosz.

Iain King – Thank you to everyone who helped hone the ideas set out in Chapter 4. There are too many names to list them all – developing quasi-utilitarianism has taken too many years – but the support of three people stands out because their generous contributions were essential: Ben Kienzle, Marianne Talbot, and David Whetham. Thank you, too, to everyone who helped me apply these ideas to violent conflict, in particular Andy Bearpark, Esther Grisnich, and Clare Hulton, the exceptional staff at the Royal College of Defence Studies (RCDS), and Heather Conley's wonderful team at the Center for Strategic and International Studies (CSIS). Finally, thank you to my family, especially Myles and Verity, for their tolerance, love, and wisdom.

ABOUT THE AUTHORS

Deane-Peter Baker [BA (Hons) Natal, MA (Natal), MSocSci (KwaZulu-Natal), PhD (Macquarie)] is Associate Professor in the School of Humanities and Social Sciences at UNSW Canberra, the branch of the University of New South Wales that is the academic service provider to the Australian Defence Force Academy (ADFA). He is Director of the UNSW Military Ethics Research Lab and Innovation Network (MERLIN) and also a co-convenor (with Professor David Kilcullen) of the UNSW Canberra Future Operations Research Group. Dr Baker came to UNSW Canberra from the US Naval Academy, where he taught in the Department of Leadership, Ethics, and Law. Prior to that, he taught in the School of Philosophy and Ethics and its predecessors at the University of KwaZulu-Natal in South Africa for over 10 years. While working in South Africa, Dr Baker served a term as Chairperson of the Ethics Society of South Africa. He has held and managed research grant funding from Australia's Department of Defence, the National Research Foundation (South Africa), and the British Academy. Currently, Senior Visiting Fellow in the Centre for Military Ethics at King's College London, Dr Baker has previously held visiting fellowships at the Triangle Institute for Security Studies (Duke University), the US Army's Strategic Studies Institute, the Centre for Applied Ethics at Stellenbosch University, and the Department of Philosophy at the University of Pretoria. In 2019, he was Senior Visiting Fellow in the Institute for Advanced Studies at Durham University.

Dr Baker works regularly with units and formations of the Australian Defence Force, especially Australia's Special Operations Command, assisting in the development of ethics-focused professional military education (PME) and ethics training efforts. He served as Panellist on the International Panel on the Regulation of Autonomous Weapons, an initiative funded by the German Federal Government. Dr Baker's Massive Open Online Course (MOOC), 'Military Ethics: An

Introduction' (FutureLearn) – the first ever MOOC on this topic – was a finalist in the 2016 ATOM Awards. Among his publications are *The Ethics of Special Ops: Raids, Recoveries, Reconnaissance and Rebels* (co-author with Roger Herbert and David Whetham, Cambridge University Press 2023); *Should We Ban Killer Robots?* (Polity Press 2022); *Morality and Ethics at War: Bridging the Gaps Between the Soldier and the State* (Bloomsbury Academic 2020); *Citizen Killings: Liberalism, State Policy and Moral Risk* (Bloomsbury Academic 2016); *Just Warriors, Inc.: The Ethics of Privatized Force* (Polity/Continuum 2010); and *New Partnerships for a New Era: Enhancing the South Africa Army's Stabilizing Role in Africa* (US Army War College 2009). Edited volumes include *The Strategic Corporal Revisited* (with David Lovell, 2017, University of Cape Town Press); *Key Concepts in Military Ethics* (2016, UNSW Press); *Private Military and Security Companies: Ethics, Policies and Civil-Military Relations* (with Andrew Alexandra and Marina Caparini, 2008, Routledge); and *Alvin Plantinga* (2007, Cambridge University Press). Dr Baker served briefly in the British Army and as a reservist in the South African Army.

Rufus Black [BA (Melbourne), LLB (Hons) (Melbourne), DipTheol (Oxford), MPhil (Oxford), DPhil (Oxford), DUniv (Victoria University)] is Vice Chancellor and President of the University of Tasmania. He has worked on questions of governance, management, ethics, and strategy in the defence and security environment at the most senior levels in Australia. As a partner of McKinsey & Company, he led the Pappas Budget Audit (2009) and subsequently conducted the Black Review into Defence Governance and Accountability (2010), co-led the Prime Minister's Independent Review into the Australia Intelligence Community (2011), and is a Panel Member of the Afghanistan Inquiry Implementation Oversight Panel (2020–).

He has taught ethical leadership to senior Australian leaders, including Defence Leaders, through the Centre for Ethical Leadership in Melbourne for nearly a decade. He has published on a range of ethical issues from bioethics and the relationship of ethics and economics through to ethical theory. At the University of Melbourne, he was Master of Ormond College, Enterprise Professor in the Department of Management and Marketing, and Principal Fellow in the Department of Philosophy. He was also Deputy Chancellor of Victoria University.

His public policy work for government has included being the Strategic Advisor to the Secretary for Education in Victoria, a member of multiple Ministerial Higher Education Reform Panels, Director of Innovation Science Australia, and a member of the Premier's Social and Economic Recovery Council (Tasmania). He is currently a member of the Australian Government's Urban Policy Forum.

Professor Black has worked extensively in the private sector. He worked for McKinsey for 9 years, where he was a partner, serving clients in Australia and Asia on organization and strategy issues, and he helped found its public sector practice in Australia. He has been Director of the national law firm Corrs, Chambers, and Westgarth and involved in the start-up sector co-founding the Wade Institute for Entrepreneurship in Melbourne.

He was also Founding Chair of the Board of Teach for Australia, Director of the Walter and Eliza Hall Institute of Medical Research, and Chair of their Human Research Ethics Committee, and he is Director Emeritus of the New York based Teach for Hall.

He was educated at the universities of Melbourne and Oxford, where he studied as Rhodes Scholar.

Roger Herbert [BS (Davidson College), MA (Naval Postgraduate School, MS (National War College), PhD (University of Virginia)] served as the General Robert T. Herres Distinguished Military Professor of Ethics at the US Naval Academy from 2018 to 2021. As the programme director for the Naval Academy's core ethics course, Dr Herbert orchestrated a fundamental restructuring and reorientation of a curriculum that has been in place since 1995 and co-edited the textbook that supports the new curriculum (Herbert and LiVecche 2023).

Dr Herbert's research focuses on the nexus between ethics and statecraft. His dissertation, *Cry Havoc: Rhetorical Mobilization and Foreign Policy Decision Making During War-threatening Crises* (2016), examines why and how statemen leverage rhetoric to marshal domestic support for their aggressive policy preferences. His current research offers an ethical critique of state propaganda during international crises. Dr Herbert's research agenda also includes an analysis of ethical challenges and concerns that are distinctive to special operations forces and the conduct of special operations. Among his publications is *The Ethics of Special Ops: Raids, Recoveries, Reconnaissance*, with Deane-Peter Baker and David Whetham (Cambridge University Press 2023).

Prior to his doctoral studies, Dr Herbert served for 26 years in the US Navy as a Naval Special Warfare (SEAL) Officer. His diverse postings included operational tours with multiple SEAL teams and staff assignments with US Naval Forces Europe, the Joint Staff, and US Special Operations Command. He commanded SEAL Delivery Vehicle Team TWO (a SEAL team that specializes in undersea operations), Naval Special Warfare Unit THREE (Naval Special Warfare's forward unit in the Middle East), and the Naval Special Warfare Center (Naval Special Warfare's training command). Dr Herbert retired from the Navy in 2010 with the rank of Captain.

Iain King is an author and defence practitioner with an extensive background in both conflict work and ethics; he spent 2022 and 2023 in Baghdad as the lead civilian advisor for NATO's Mission in Iraq. Previously, he led international civilian operations in Benghazi during the Libyan civil war, pioneered local stabilization operations in Afghanistan's Helmand province, held key posts with the EU and

UN in Kosovo, and spent 7 years working on the Northern Ireland peace process. He also directed the UK's government research programme into conflict and was a Fellow at the US-based think tank, the Centre for Strategic and International Studies (CSIS), where he wrote multiple articles on defence and security, including for NBC, as well as being an in-studio commentator on the Balkans for CNN and BBC. Iain was made a Commander of the British Empire (CBE) for his work abroad in 2013 and a Fellow of the Royal Society of Arts (FRSA) for his literary and academic work in 2015.

Iain has written five books. His first, on Kosovo, was described by an academic journal as *'The most perceptive account ever written . . . of peace-building in the aftermath of ethnic conflict'* and by the Economist magazine as *'Excellent and well written'*. Another, on Afghanistan, was acclaimed by the *British Army Review* as *'The best written account of the Helmand campaign'*. One of his two novels became Britain's best-selling spy story of the year.

His major work on ethics was the satirically titled, 'How to Make Good Decisions and Be Right All the Time'. Published in 2008, this set out Iain's theory of ethics, which commentators have described as 'quasi-utilitarian'. He has also written about ethics for the popular UK magazine *Philosophy Now*, with articles such as *'Moral Laws of the Jungle'* and *'Should Indiana Jones Eat the Monkey Brain?'* His twin interests of war and philosophy came together with his *Thinkers at War* series for *Modern History Magazine*, which explored how famous philosophers, from Socrates to John Rawls, were influenced by their wartime experiences.

Iain holds master's degrees from Oxford University and Kings College London, was Research Fellow at Cambridge University, and has been Adjunct Scholar at the US Modern War Institute at West Point.

1

INTRODUCTION

Deane-Peter Baker, Rufus Black, Roger Herbert, and Iain King

Heavy and sustained fire from a well-fortified machine-gun position on high ground has effectively pinned down your patrol. One of your soldiers, Corporal Briggs, is severely wounded and bleeding heavily. Your medic is doing what he can for her, but he fears she won't survive unless she's evacuated asap to a higher level of medical care. Your team has made several attempts to get her out of the kill zone, but that gun sees your every movement and seems to have a limitless supply of ammo. Your attempts to silence the gun have also been ineffective, and the fire mission you've just called in may be hours away.

'Are you sure you can't take out that shooter?'

'I'm sure', your sniper, Sergeant Freeney, answers, 'not at this range. He's just too dug in'. He then points toward a road that leads down a hill from the insurgent's firing position. 'But *they* aren't'.

You follow your sniper's index finger to three vehicles at the base of the hill. 'See those kids by the white pickup?' asks Freeney. You see a girl, maybe 12 years old, just a few years older than your own daughter. She looks terrified. Two boys then appear, about the same age as the girl, maybe younger. 'They've been resupplying that gun, sir. Take them out, and that gun will be out of business in minutes'.

You watch the girl collect ammo cans and realize Freeney's right. It seems you have two bad choices. If you hunker down and wait for the fire mission, Briggs may not survive; she's already looking ashen from blood loss. Alternatively, you can order Freeney to kill the children – or at least one of them to scare the others away – and move out as soon as the Taliban gun runs dry. Given his rate of fire, it won't take long.

As hard as this decision is for you, you think about Freeney. He's a tough soldier who will do what he has to do, but Freeney has kids too. You'd be ordering him

DOI: 10.4324/9781003312925-1

to steady his crosshairs on a child, pull his trigger, and watch a plume of red and a crumpling child.

What are you going to do, Lieutenant? 'Quickly, sir', shouts your medic. 'I don't think Corporal Briggs has much time'.

We've dramatized this vignette and added fictitious names, but it's no thought experiment. An Australian soldier shared this story with one of our book's co-authors. Although the ethical challenges that soldiers[1] face aren't always so harrowing, moral complexity is an inherent feature of military life. And because all states, by most definitions of what constitutes a state, require a segment of the citizenry to be always ready to violently defend the state's political sovereignty and territorial integrity, governments incur a moral obligation to prepare the young men and women who volunteer for (or are conscripted into) the armed forces of the state to manage the exceptional moral burden of military service. Even for states that eschew the language of morality in matters of national security, the practical incentives for attending to the moral resilience of those who serve are compelling. Upon coming home, soldiers who have been morally injured by war become burdens to the societies they once served.

States have long understood the imperative to prepare their armies physically and materially for war. Over the past century, most states have also made impressive advances toward acknowledging the need to prepare their troops psychologically. Recognition of the state's responsibility to 'morally armour'[2] their soldiers, to prepare them for the ethical trials of war, has lagged severely, even in the liberal democratic states that are the focus of this book. Indeed, until Jonathan Shay's influential *Achilles in Vietnam* (1994), moral injury (a response to violations of deeply held *moral* commitments) was not widely accepted as a pathology distinct from posttraumatic stress disorder (a response to *psychological* trauma). While Shay's contribution has inspired an impressive body of scholarly research, few states invest sufficient time or resources in the moral readiness of their forces.

So, what *should* states be doing to prepare their service members for ethical challenges as prickly as the one summarized in the aforementioned vignette?

First, members of the armed forces require at least an elementary grasp of the Laws of Armed Conflict (LOAC)[3] and, when sent into action, a *detailed* understanding of the rules of engagement (ROE) that translate legal principles into explicit guidelines for the military operation at hand. By acting within the confines of the LOAC and a set of ROE, soldiers are more likely to make sound ethical decisions. However, an action can be both legal and *immoral*. What is legal and what is ethical are closely related questions. Ideally, the answers overlap significantly. But law and ethics are distinct disciplines that ask fundamentally different questions. Laws establish what military personnel *may* (and may not) do. However, the law is often mute regarding what service members *ought* (and ought not) to do. 'Ought' is the purview of ethics. In the previous example, the patrol leader must choose between two legal options. He is neither legally constrained from killing the children nor is he legally obliged to do so. Because the children are actively involved in the sequence of events threatening his patrol, they have tragically acquired the legal status of combatants. Both of your awful options, therefore, are legal, and neither can be said to be *more legal* than the other. The law doesn't work that way.

Given the limits of the law in addressing ethical questions, it seems states also owe their military personnel basic training and education in military ethics. The criteria of *jus in bello*, that part of the just war tradition that addresses ethical conduct in war, provide essential ethical guidance regarding who on the battlefield a soldier may justly kill and how they may kill them. The *jus in bello* convention asserts that only combatants may be intentionally targeted (the principle of *discrimination*), that collateral damage – foreseeable though unintentional harm – is only acceptable to the extent that it's proportional to the value of the target being attacked (the principle of *proportionality*), and that combatants employ the least harmful means necessary to achieve their legitimate military ends (the principle of *necessity*).

As important as the *jus in bello* principles of discrimination, proportionality, and necessity are for soldiers negotiating complex ethical terrain, they are still not sufficient. For one thing, the *jus in bello* convention only addresses those ethical challenges that involve the use of force. However, the moral complexity of military life is not limited to kill-or-don't-kill decisions. Nor do the *jus in bello* criteria resolve every situation that *does* involve the use of force. For example, in the scenario outlined earlier, considerations of discrimination, proportionality, and necessity offer the patrol leader little help in determining the right thing to do. Both options earn passing marks for proportionality and necessity. And while deliberately targeting children will typically fail (badly) to meet the standards of discrimination, it does not in this case. Because the children supplying the machine gunner are 'engaged in the business of war' (Walzer 2000, 43), they have forfeited noncombatant immunity. Better put, those who wrongfully directed the children to perform a military task forced them into a rights forfeiture.

Because the *jus in bello* convention targets only a subset of the moral challenges military personnel are likely to face in the conduct of their duties, we argue that more substantial moral armouring is required for a state to meet its moral duty to its service members. An implied task in the state's obligation to train and educate members of its armed forces in the essentials of military ethics is to train and educate them in the fundamentals of *moral reasoning*. We further maintain that the state cannot fulfil this duty merely by instituting a mandatory survey course in moral philosophy; soldiers in the field need a process for putting esoteric ideas to work for them in the real world. Without a framework for applying theory to practice, basic familiarity with moral theory cannot achieve the purpose behind the obligation: to arm soldiers with a capability for rapid moral deliberation under conditions of physical discomfort, uncertainty, and danger. We argue, therefore, that it is essential for states to equip their soldiers, sailors, marines, and aviators with a repeatable and reliable ethical decision-making framework that will enable them to address ethical challenges in a consistent, appropriate, and, when necessary, expeditious way.

The imperative to provide service members with a deliberative framework to discipline their moral reasoning prompts the question at the heart of this project: what framework should the armed forces of liberal democratic states adopt? In this volume, four authors – each with experience and credentials in the field of military ethics – attempt to answer this question. Deane-Peter Baker presents an ethical decision-making model he calls Ethical Triangulation, a model he has employed throughout his career as an applied military ethicist, including in his current role educating Trainee Officers of the Australian Defence Force Academy. Roger Herbert, a retired US Navy SEAL and former Robert T. Herres Distinguished Military Professor of Ethics at the US Naval Academy, outlines the Moral Deliberation Roadmap, the deliberative framework formally adopted by the US Naval Academy in 2021. British writer Iain King, a scholar at the Modern War Institute at the US Military Academy at West Point, presents a hybrid methodology for ethical decision-making – honed in conflict-affected states, including Kosovo, Libya, Afghanistan, and Iraq – that has been described as quasi-utilitarian. Finally, Rufus Black, Vice Chancellor of the University of Tasmania, advocates for an approach to moral reasoning based in natural law, one which has heavily influenced the ethical decision-making framework adopted by the Australian Defence Force in 2021.

In Part I of our book (Chapters 2–5), each author outlines his favoured ethical decision-making approach. In Part II (Chapters 6–9), we each offer a critique of our co-authors' contributions, highlighting both merits and deficiencies. The book concludes with an Afterword from our colleague David Whetham, Professor of Ethics in the Defence Studies Department of King's College London, which encapsulates the debate, summarizes areas of agreement between the authors, and identifies critical areas of difference.

Our Dragons

Academic writing should not only contribute fresh ideas but also attempt to slay a dragon or two. It should address competing theories or make a case that current thinking on a given topic is incomplete, deficient, or just plain wrong. For this project, it may appear that our dragons are built into the book's structure; our text is organized as a debate among the four authors.

While it's true that our critique chapters pull no punches, they don't quite rise to the dragon-slaying level. The reason we say this is that our debate is essentially an intramural affair. The four authors contributing to this volume may disagree on how best to design an ethics curriculum for members of the armed forces, but we agree on first principles, the fundamental concepts, and basic assumptions of military ethics. For example, we reject the assertion advanced by some political realists that war is an *amoral* enterprise, a category of human endeavour that lies beyond the scope of ethical critique, neither moral nor immoral. Likewise, *contra* pacifist avowals, we agree that some injustices warrant the resort to state violence. We agree, in other words, that war is not categorically immoral.

Another basic assumption that the authors of this book hold in common is that moral theory can and should shape human behaviour. This assumption, however, is by no means a given. There are two theoretical frameworks – situationism and intuitionism – that challenge this postulation. Neither research programme has much to say regarding whether moral theory *should* inform behaviour, but both are deeply sceptical about the extent to which it *can*. These are the dragons at which our project must tilt.

Situationism

Situationism posits that character and rational analysis are peripheral to the moral judgements we make. Our behaviour, situationists maintain, is primarily determined by situational factors and social context. Drawing on an impressive body of experimental research, situationists observe that even 'unobtrusive features of situations . . . impact behavior in seemingly arbitrary, and sometimes alarming, ways' (Doris *et al*. 2020). For example, few of us imagine ourselves capable of administering potentially lethal shocks to fellow humans. Yet for most participants in Stanley Milgram's 1963 experiment, authoritative instructions from the laboratory-coat-clad experimenter proved sufficient to overcome moral reservations. In 1971, psychologist Philip Zimbardo was forced to cut short his experiment when Stanford University students who were directed to role-play prison guards became disturbingly abusive to fellow students role-playing their prisoners. In another well-known experiment, subjects who found a coin in a payphone (back when that was a thing) were 22 times more likely to offer aid to a role-player fumbling with a stack of papers than those whose mood had not been lifted by the unexpected coin (Isen and Levin 1972). In a similar experiment, Darley and Batson (1973) found that unhurried Princeton University seminary students were six times more likely to stop and offer aid to role-players in clear physical distress than those who were in a rush to present their papers at a venue across campus.

In the narrative that introduces this chapter, the primary situational influencer confounding the patrol leader's decision is not at all subtle or 'unobtrusive' like a coin in a pay phone. Nevertheless, it demonstrates well the situationist claim. Absent a situational influencer, the patrol leader's decision isn't particularly difficult. The ammo haulers have, through their actions, acquired the legal and moral status of combatants. Killing them would be both lawful and, arguably, morally permissible. It would be a stretch to say that targeting the ammo haulers is morally *required* in this case, but if the patrol leader fails to give that order, and Corporal Briggs bleeds out while waiting for the fire mission, explaining that decision to Briggs' parents would be an unenviable task. The only reason this decision is difficult – indeed, gut-wrenchingly difficult – is that the ammo haulers are children. Soldiers protect children. They don't put children in their crosshairs.

Professor Dennis Vincent, who leads the Department of Communication and Applied Behavioural Science at the Royal Military Academy Sandhurst (RMAS), takes seriously 'the power of the situation' to affect moral judgement. After summarizing an extensive literature, Vincent (2022, 9–17) argues that situational influencers (e.g. a hostile environment, normalized violence, weak leadership, a lack of resources) tend to lead to common behaviours and cognitive biases that diminish a soldier's capacity for intentional moral deliberation. Vincent summarizes ten common behaviours, which he usefully breaks down into individual behaviours (conformity, de-individuation, obedience, cognitive dissonance, bystander effect) and group behaviours (groupthink, risky shift, authority, dehumanization, and status quo bias). These common behaviours, Vincent convincingly argues, can result in moral judgements that are radically out of synch with the decision-maker's character.

Sandhurst has responded to the situationist challenge by taking positive measures to account for and, ideally, neutralize the outsized effect of situational influencers that situationist researchers describe. The 2014 RMAS Leadership Conference committed to the development of 'a simple toolkit for remembering the key principles of Ethical Leadership' (Ibid., 25). The result is the Vincent S-CALM Model (Situational Influencers, Common Behaviours, Accountability, Leadership, Moral Compass), which is currently taught to all Officer Cadets at RMAS. While training and education can never fully prepare service members for situational contingencies like the appearance of child soldiers on the battlefield, Sandhurst leadership believes that officers trained in the Vincent S-CALM Model will be better prepared to recognize situational influencers and the common behaviours they tend to produce, thereby overcoming or at least mitigating their ethically corrosive effects.

Intuitionism

The second dragon that calls into question the value of our project is *moral intuitionism*. Like situationism, intuitionism holds that rational analysis plays a humbler role in ethical decision-making than most moral theorists care to admit. However, unlike situationists who focus on the influence of external factors, intuitionists look internally to the psychology of moral judgement and the nature of moral knowledge itself. For intuitionists, it's not the situational context that sometimes interferes with our capacity to reason deliberately, it's the essential cognitive inclination of our species. We're wired, intuitionists say, to perceive the moral world through our intuitive faculties, not our conscious intellect.

Intuitionism is a variegated research programme that traces its heritage to the ancients and incorporates insights from multiple disciplines including philosophy, psychology, sociology, evolutionary biology, and neuroscience. A comprehensive account, therefore, is beyond the scope of this chapter. Instead, we focus on moral psychologist Jonathan Haidt's conceptualization presented in his influential 2012 book *The Righteous Mind: Why Good People Are Divided by Politics and Religion*. Because Haidt presents his 'social intuitionist model' in such a cogent and accessible way (it's a must-read!), it has earned the attention of a broad audience, not just academics.

Haidt draws on research from experimental psychology, sociology, and evolutionary biology to make a case against the 'rationalist delusion' that our moral judgements are the result of deliberate, evidence-based reasoning (Haidt 2012, 28). 'From Plato through Kant and Kohlberg', he writes, 'rationalists have asserted that the ability to reason about ethical issues *causes* good behavior' (Ibid., 89). Haidt rejects this and asserts as his 'first principle of moral psychology' that 'intuitions come first, strategic reasoning second' (Ibid., 52). Well before our rational minds can start consciously noodling about how Kant, Aristotle, or our favourite ethics professor might work through a given ethical problem, our moral intuitions have already decided what to do 'automatically and almost instantaneously' (Ibid., xiv). If we engage in deliberate moral reflection at all, we do so in search of *post hoc* rationalizations to justify decisions we've already made (Ibid., 26). Although he believes David Hume's 'master-slave' analogy overstates the disparity in status between intuition and reason, his own metaphors (e.g. elephant-rider, client-lawyer) make it clear that intuition is the boss. Reason, as he characterizes it in another helpful analogy, is merely the 'press secretary who automatically justifies any position taken by the president' (Ibid., 91).

If Haidt's model is right, there would be little reason for the reader to proceed. At best, the four models we present in the chapters ahead would offer little more than clever strategies for Haidt's 'press secretary' to justify gut intuition.

Haidt convincingly argues his first principle of moral psychology, that intuition comes first in the chronology of ethical decision-making and fortifies this claim with compelling experimental evidence. But it really needs no defence. Most of us simply *know* this to be true from our personal experiences as moral agents. We're all 'cognitive misers' who seek to conserve mental effort whenever we can (Fiske and Shelley 1991). We're less convinced, however, of Haidt's insistence on the *stickiness* of pre-reflective moral judgement. While it makes evolutionary sense that we should pre-reflectively formulate courses of action to 'triggering events' (Ibid., 47) and act on those formulations without further reflection if time for deliberation is scarce, we typically *do* have time for reflection and, importantly, for revision of our intuitive judgements when we discover 'cues that an intuitive judgement could be wrong' (Kahneman and Klein 2009, 519).

Haidt generously offers experimental evidence that supports our criticism of his model (Ibid., 69–70). Researchers Joe Paxton *et al.* (2012) forced participants in the treatment group of their experiment to wait 2 minutes before answering survey questions about a morally complex scenario. The researchers found that the period of forced reflection significantly influenced the substance and nuance of participant responses. Drawing on his elephant-rider metaphor, Haidt (Ibid., 70) concedes that 'if you force [the elephant and the rider] to sit around and chat for a few minutes, the elephant actually opens up to advice from the rider and arguments from outside sources'. Nevertheless, he maintains that this just doesn't happen very often.[4] Without an artificial mechanism to force deliberation, he argues, people make their moral judgements quickly and emotionally regardless of the time available for reflection. So, while it's 'possible for people simply to reason their way to a moral conclusion that contradicts their initial intuitive judgement', he insists that this is so rare that it doesn't warrant complicating the social intuitionist model with additional variables (Ibid., 67).

As a descriptive claim, it may be true that we humans are indeed wired to decide with our guts before consulting our brains. Yet for the purpose of our project, what matters is not whether people tend to disregard reason and act on intuition when making moral choices, but whether people – particularly people wearing military uniforms – are capable of incorporating rational deliberation into their ethical decision-making strategies. Outside of the laboratory, we see that people are rather adept at suppressing gut responses. Indeed, some professions *require* their members to engage in behaviours that not only contradict intuition but also the very instinct for self-preservation. People intuitively flee their homes in the event of an uncontrollable house fire. They may pause to ensure their loved ones are accounted for, gather up the cat and the goldfish, and grab the laptop that hasn't been backed up in ages. After that, they're gone. Presumably, firefighters, as humans, are similarly wired to run away from fire. Yet as we civilians flee, firefighters run in the opposite direction. We admire their courage, but we also *expect* them to behave in a manner at variance with intuition and instinct. It's their job.

For obvious reasons, firefighters make good proxies for soldiers when studying decision-making processes. It's reasonable, therefore, to extrapolate from Klein *et al.*'s 1986 analysis of how fireground commanders make time-critical decisions in the heat of 'battle' (see also Klein *et al.* 2010). The researchers found, as Haidt would predict, that intuition comes first. Relying on the repertoire of patterns compiled through years of firefighting experience, fireground commanders intuitively settled on a candidate course of action without necessarily being aware of the cues that led to that choice. But intuition is not the final arbiter. Despite the inherent urgency and physical risks of firefighting, fireground commanders deliberately run their intuitive judgements – course of action A – through a mental simulation process to estimate the likelihood of success. If COA A seems plausible, it's implemented without considering other options. If, however, the simulation reveals potential problems, fireground commanders turn to 'the next most available and/or similar option' and repeat mental simulations of each candidate COA until one COA emerges that survives scrutiny (Klein *et al.* 1986, 579). Although the key finding of the study was the serial nature of COA analysis – the researchers had hypothesized pairwise comparison strategies – the study also suggests that for expert decision-makers, the relationship between intuition and rationality is a partnership. Rationality relies on intuition to serve up options, but ultimately, it's the rider, not the elephant, calling the shots.

Narvaez (2008) artfully describes the relationship between intuition and reason as a dance, with each impulse 'taking turns doing the leading' in an iterative process that is integrated (a nod to situationism) 'with the history, needs, and goals of local circumstances'. But, even if we concede that people generally lack the *proclivity* to engage in rational deliberation prior to moral judgement as the social intuitionist model posits, we must at least acknowledge that we have the *capacity* for deliberate moral reasoning. What is needed to nurture this capacity is that external mechanism to which Haidt alludes. In the Paxton, Unger, and Greene experiment, that mechanism was a timer that prevented the treatment group from registering their moral judgements for 2 minutes. For soldiers, as with firefighters, the mechanism is extensive experience, training, and education in decision-making under stress, reinforced by leaders who rely on a quiver of sticks and carrots to disincentivize impulsiveness and promote intentional and structured moral deliberation.

Before judging us Pollyannaish for believing that an ethics class and a drill sergeant can train away millennia of evolved behaviour, consider this. When poorly trained infantry units take enemy fire while on patrol, they hastily return fire. By contrast, the immediate reaction of well-trained troops to enemy fire is to drop and think. Where is the fire coming from? Is it effective fire? Does the enemy know our position and strength, or could the enemy be trying to draw us out by conducting a 'recon by fire'? Where are the exits? Where is the nearest cover and concealment? Are there friendlies in the area? Could this be friendly fire? Experienced patrol leaders may be capable of weighing these considerations in seconds, but regardless of how long it takes, they resist the impulse to act intuitively. The first principle of infantry tactics, one might say, flips Haidt's first principle on its head: strategic thinking comes first; intuitions are for amateurs. If soldiers can learn to stop and deliberate while under fire, it seems we can teach them to count to 120 when confronting a moral dilemma.

Haidt also underestimates the role that reason plays in the aftermath of a decision. While he acknowledges that post-judgement self-reflection, if approached sincerely, can increase our moral reasoning aptitude, he chooses to demote the importance of self-reflection by assigning it a dotted line in his schema.[5] He offers two reasons for the demotion. First, he believes that we aren't very good at self-reflection and, second, that it 'doesn't seem to happen very often' (Ibid., 47). But these conclusions are far from self-evident. The moral injury literature, for example, generally supports his first claim, that we're not very good at self-analysis. It's common for self-reflection of battlefield decisions to result in erroneous conclusions that contribute to moral injury. A typical case involves a soldier blaming herself for circumstances that were obviously beyond her control (see especially Sherman 2015). However, his second conclusion, that we don't often engage in private reflection on our moral choices, is demonstrably wrong. Indeed, for many victims of moral injury, that's the problem. The literature abounds with cases in which service members simply cannot stop reflecting on the choices they made, regardless of how unhelpful or unhealthy this reflection may be.

Finally, while Haidt undervalues the role that reason plays in post-judgement analysis, he seems to reject outright the proposition that reason can serve a *pre-judgement* function. 'Nobody is ever going to invent an ethics class that makes people behave ethically' (Haidt 2012, 90). We agree that experience, good and, especially, bad, followed by reflection is unquestionably the best way to refine moral intuition and sharpen moral reasoning skills. However, as unapologetic rationalists, we are steadfast in our belief that excellent ethics curricula, especially curricula that are rich in realistic case studies, not only instruct and enhance a student's moral intuition but also increase the likelihood that reason will be brought to bear when our students confront moral dilemmas in the wild. Retired US Navy Captain Rick Rubel, the US Naval Academy's Robert T. Herres Distinguished Military Professor of Ethics from 1995 to 2018, frequently shares with his students and colleagues a vignette that is illustrative of the role that moral theory and a good ethics course can play in shaping moral intuition and the capacity to make difficult moral choices when time is short and the stress-meter is pegged. One of Captain Rubel's students, a future Marine, confronted the Captain after class and asked, 'Do you really expect us to think about all this ethics stuff every time we're about to pull the trigger on a battlefield?' 'No', replied Captain Rubel, 'I expect you to think about all this ethics stuff now so that on the battlefield, you won't have to'.

Situationism and intuitionism are productive research agendas that have generated important insights and will continue to do so. Both theoretical frameworks, however, consign reason – and the likes of Aristotle, Thomas Aquinas, Kant, and Mill – to the back bench. In doing so, they promote a form of moral defeatism that we reject. In the case study at the beginning of this chapter, an understanding of situationism and intuitionism may alert the patrol leader to potential blind spots of moral perception. Beyond this, however, they are not helpful. They offer him no advice regarding which course of action he should take. A still more pernicious effect of the situationist and intuitionist verdicts is that they provide the patrol leaders' superiors with justifications, backed by the imprimatur of science, for not investing scarce resources (time and money) into the ethical readiness of the patrol leader and his troops. If situational influencers that cannot be anticipated ultimately determine the patrol leader's decisions, wouldn't his limited training time be better spent on the rifle range than in an ethics class? If the patrol leader is cognitively wired to rely on gut intuition regardless of his training and education he's received, why invest in standing up ethics curricula for officer candidates and senior NCOs? In short, situationism and intuitionism invite us to shrug at the idea that moral theory can or should play a role in moral decision-making. This is a dangerous ethical orientation under any circumstances, but especially so on the battlefield, where moral defeatism can end in atrocity.

Since it's unlikely that these formidable dragons will yield to our humble pens, it's important to bear in mind that a commitment to situationism or intuitionism does not rule out embracing one or all four of the deliberative frameworks advanced in the ensuing chapters. Situationism and intuitionism are principally

descriptive, not normative, research programmes. As Haidt assiduously emphasizes, his social intuitionist model theorizes 'how the moral mind *actually* works,
not how it *ought* to work' (Ibid., 120). In other words, it's possible to hold simultaneously and without contradiction, the dual convictions that we ordinary humans
tend to make moral choices quickly, emotionally, and unmindfully, *and* that members of the armed forces (and similar professions) ought not to do so.

In the following section, we provide a rationale for the latter conviction.

The Exceptional Moral Burden of the Profession of Arms

For some professions – for example, medicine, law, and the clergy – a generally
sound moral intuition is necessary but insufficient. We expect these professionals
to possess a refined ability to recognize and negotiate the complex ethical challenges inherent in their fields. In the introduction to the textbook for the US Naval
Academy's core ethics course, Herbert and LiVecche (2023) argue that the profession of arms is among those occupations that impose exceptional ethical demands.
So, while the moral mind may actually work on intuition, as Haidt so effectively
argues, the moral minds of the people we deploy in the service of our societies
ought to function on a different level.

The authors support this assertion by suggesting seven qualities inherent in
the profession of arms that distinguish the moral burdens of soldiers from those
of other disciplines. As they are equally applicable to this project, we offer a
summary:

1 Military service is a violent enterprise. 'Fighting and winning wars', the authors
   observe, 'involve killing on a large scale, which, regardless of the righteousness
   of the cause, will always be morally problematic'. Moral intuition and judgement may well precede moral reasoning, but if that judgement involves killing
   at scale, the press secretary (to continue using Haidt's analogy) had better be
   damn good. Mere rationalizations are insufficient warrants for taking human
   lives. Killing demands moral justification, not excuses. If the press secretary
   can't muster one, the soldier must modify his/her initial judgement or suffer
   moral, psychological, or legal consequences.
2 Combat is a radically novel activity. Experience shapes moral intuition. If we're
   fortunate, parents, grandparents, teachers, preachers, and coaches also contribute. For most of us, this is adequate for negotiating the quotidian moral challenges of life. Military service, however, 'imposes unique moral demands . . .
   [that] are intensified for those in leadership positions and further amplified for
   those who lead under the stress of combat'. In short, most service members step
   onto the battlefield with no ethical reference points; it's an alien world, morally
   speaking, from the one in which their intuitions evolved. 'Any moment may
   present a soldier with a moral problem that the "folks back home" could never
   have conceived and will never fully understand'.

3 Military service involves the moral burden of leading others. 'Leadership responsibility', write Herbert and LiVecche, 'compounds the moral burden'. Regardless of rank, most military personnel will, at some point, need to make a moral choice that others must follow. If the decision-maker misjudges, he does not suffer alone. 'Most leaders', the authors continue, 'feel a moral obligation to do their best to bring their troops home alive. However, bringing them home physically whole is not the end of that discussion. Leaders must also strive to bring their troops home as whole human beings'.

4 The military profession amplifies the influence of time on decision-making. Although often, even in war, we tend to exaggerate urgency, military decision-making is routinely time sensitive. Given the unreliability of intuition in the radically novel moral circumstances of war, service members need a script, 'a well-rehearsed process or framework for rapidly working through the salient factors of a moral choice'. Without such a framework, a soldier is at risk of moral failure 'not because he is a bad person, but because he failed to ask the right questions of a given situation'.

5 The standards of ethical behaviour, by tradition and for the reasons listed earlier, are exceptionally demanding in the military cultures of most liberal democratic states. There is typically some degree of ethical culture shock for those who volunteer for or are conscripted into the military service of these states. They quickly discover that many behaviours considered tolerable when they were ordinary citizens are aggressively disciplined by their military chains of command. For example, our societies may insist that lying is wrong, but liars are rarely held to account, and some are handsomely rewarded. We roll our eyes and shrug our shoulders. But lying, cheating, or stealing is incompatible with a culture that depends on trust. 'A pilot would never step into a cockpit if she feared that her aircrew had lied about conducting critical maintenance on her aircraft or cheated on the exam that qualified them to conduct that maintenance or planned to rifle through her belongings while she is flying'. Recruits must adapt quickly to the military's culture of 'rigorous truthfulness' or risk never recovering the trust of their chains of command or their peers (Lucas 2014, 4).

6 The moral obligations of military service are frequently in competition. For military personnel, general obligations, the duties we owe to all humans, as humans, are routinely at odds with special obligations, duties that result from bonds of loyalty to mates forged by mutual commitments and sacrifices. Not as routinely, but also not unheard of, are situations in which obligations to follow lawful orders are at odds with deep-seated personal moral commitments. It is the nature of a moral dilemma that fulfilment of one moral commitment requires setting aside another. Service members must not only be adept at working through routine encounters with moral dilemmas but also make judgements that fellow soldiers can reasonably anticipate.

7 In liberal democratic states, the military's efficacy and legitimacy depend on the trust of the citizenry. In a letter to a subordinate general, General George Marshall observed, '[We] have a great asset, and that is our people, our countrymen, do not distrust us and do not fear us' (quoted in Neustadt and May 1988, 247). Without public support, recruiting suffers, funding dries up, capabilities erode, and morale tanks. 'I don't want you to do anything', Marshall continued, 'to damage this high regard in which the professional soldiers in the Army are held by our people'. Marshall understood that in liberal societies, it is not enough that armies fight and win their nations' wars. They must do so in a manner that reflects the ideals of the societies they serve. Failure risks losing the 'great asset' of public support (Ibid.).

Let's Debate This

In this brief introduction, we set out to outline the purpose and strategy of our book and summarize theoretical frameworks that challenge the fundamental premise of our project. We offered several critiques, but our essential claim against situationism and intuitionism is that even if these theories accurately describe how most people make moral choices, military personnel are not most people. There are qualities inherent in the profession of arms that require members of that profession to approach decision-making more deliberately than the average citizen.

A committed situationist or intuitionist might respond that this is too big of an ask. Expecting newly reporting service members to revise how they think and make decisions under stress is to sign on to Haidt's 'rationalist delusion' (Haidt 2012, 28). We disagree and turn again to intuitionist Jonathan Haidt as the source of our rationalist optimism. Haidt is an admirer (or at least a sympathizer) of Glaucon, Socrates' student who appears in Book II of Plato's *Republic* to argue that people behave morally not because they admire justice but because they fear social disapproval and loss of reputation. Although Plato gives Socrates the upper hand in this exchange, the field of moral psychology has done much to rehabilitate Glaucon. Haidt thinks Glaucon got it right. 'We are obsessively concerned about what others think about us', he writes (Ibid., 91). So, if the goal is designing a community where members strictly adhere to an ambitious ethical code, 'the most important principle . . . is to make sure that everyone's reputation is on the line all the time' (Ibid., 74).

Haidt has unintentionally described every regimental mess – or platoon bay or ready room or NCO club – from Homer's Iliad to the present day. The military, sometimes for harm but typically for good, is an honour (and shame) culture. 'It's better to die than to look bad' is a sentiment shared among soldiers in jest, but it doesn't miss the mark by far. War crimes are predictable outcomes for military cultures that attach honour to brutality on the battlefield. As we write this book, the documented war crimes committed by undertrained and unprepared Russian soldiers in Ukraine have accumulated staggeringly. By contrast, in military cultures that attach honour to just warfighting and to the capacity to reason deliberately despite the danger, fatigue, and uncertainty of combat, you will find people inspired to make the cognitive leap – if such a leap is necessary – to commit to a repeatable and reliable ethical decision-making model.

What follows is a debate over the form that the model should take.

Notes

1 Although sailors, airmen, and marines may bristle, we use the term 'soldier' throughout this text as shorthand for any uniformed member of the armed forces.
2 'Moral armouring' is an expression we borrow from our colleague, David Whetham. See Baker *et al.* (2023).
3 Also known as International Humanitarian Law (IHL).
4 Frederick (2005) supports Haidt's observation. Frederick concludes not only that we tend to make decisions intuitively but also that the reliability of our intuition does not seem to justify our confidence in it.
5 Haidt notes that his dotted-line demotion is 'One of the most common criticisms of the social intuitionist model from philosophers' (324, endnote 44).

References

Baker, Deane-Peter, Roger Herbert, and David Whetham. 2023. *The Ethics of Special Ops: Raids, Recoveries, Reconnaissance, and Rebels*. Cambridge University Press.

Darley, John M., and Daniel C. Batson. 1973. '"From Jerusalem to Jericho": A Study of Situational and Dispositional Variables in Helping Behavior.' *Journal of Personality and Social Psychology* 27(1), 100.

Doris, John, Stephen Stitch, Jonathan Phillips, and Lachlan Walmsley. 2020. 'Moral Psychology: Empirical Approaches' in Edward N. Zalta (ed.) *The Stanford Encyclopedia of Philosophy*. Metaphysics Research Lab, Stanford University.

Fiske, Susan T., and Shelley E. Taylor. 1991. *Social Cognition*. McGraw-Hill Book Company.

Frederick, Shane. 2005. 'Cognitive Reflection and Decision Making.' *Journal of Economic perspectives* 19(4), 25–42.

Haidt, Jonathan. 2012. *The Righteous Mind: Why Good People Are Divided by Politics and Religion*. Vintage.

Herbert, Roger, and Marc LiVecche. 2023. 'Introduction' in Roger Herbert and Marc LiVecche (eds) *Ethics and Moral Reasoning for Naval Leaders*. Pearson Education Inc.

Isen, Alice M., and Paula F. Levin. 1972. 'Effect of Feeling Good on Helping: Cookies and Kindness.' *Journal of Personality and Social Psychology* 21(3), 384.

Kahneman, Daniel, and Gary Klein. 2009. 'Conditions for Intuitive Expertise: A Failure to Disagree.' *American Psychologist* 64(6), 515–526.

Klein, Gary, Roberta Calderwood, and Anne Clinton-Cirocco. 1986. 'Rapid Decision Making on the Fire Ground.' *Proceedings of the Human Factors Society Annual Meeting* 30(6), 576–580.

Klein, Gary, Roberta Calderwood, and Anne Clinton-Cirocco. 2010. 'Rapid Decision Making on the Fire Ground: The Original Study Plus a Postscript.' *Journal of Cognitive Engineering and Decision Making* 4(3), 209.

Lucas, George R. 2014. 'Ethics and the Military in America' in Rick Rubel and George R. Lucas (eds) *Ethics and the Military Profession: The Moral Foundations of Leadership*. Pearson.

Narvaez, Darcia. 2008. 'The Social Intuitionist Model: Some Counter-Intuitions' in W. Sinnott-Armstrong (ed.) *Moral Psychology, Vol. 2. The Cognitive Science of Morality: Intuition and Diversity,* Boston Review, 233–240.

Neustadt, Richard, and Ernest R. May. 1988. *Thinking in Time: The Uses of History for Decision-Makers*. Free Press.

Paxton, Joseph M., Leo Ungar, and Joshua D. Greene. 2012. 'Reflection and Reasoning in Moral Judgment.' *Cognitive Science* 36(1), 163–177.

Shay, Jonathan. 1994. *Achilles in Vietnam: Combat Trauma and the Undoing of Character*. Scribner.

Sherman, Nancy. 2015. *Afterwar: Healing the Moral Wounds of Our Soldiers*. Oxford University Press.

Vincent, Dennis. 2022. *The Vincent S-CALM Model*. Royal Military Academy Sandhurst.

Walzer, Michael. 2000. *Just and Unjust Wars*, 3rd edition. Hachette UK.

PART I

Theory

2

THE ETHICAL TRIANGULATION MODEL

Deane-Peter Baker

Imagine yourself in the following scenario:

You are leading a patrol tasked with a time-sensitive raid on a home in which a suspected mid-tier insurgent leader is believed to be holding a meeting. *En route* to the target, you pass a group of local men who are beating a woman. Your interpreter informs you that she has been accused of witchcraft and is at risk of being killed by the mob. You can

1 intervene to stop the attack and potentially miss the opportunity to complete your mission,
2 continue on your mission and accept that the woman may be killed, or
3 split your patrol and try to achieve both goals, but at the cost of significantly increasing the risk to the members of your team.

This scenario (derived from a firsthand account of a real-life operational situation) is just one example of the ethically challenging decisions that military personnel face. Ethical challenges range from routine to complex and can occur in barracks, during training, and on operations. Law provides the parameters of what military personnel *may* do but is often mute on what they *ought* to do. Consequently, military forces in liberal democracies are increasingly recognizing the need to equip their uniformed personnel with a repeatable and reliable military ethical decision-making method that will enable them to address these ethical challenges consistently and appropriately. But what method should be selected?

The principles of the *jus in bello*, the part of the just war tradition which addresses the ethical use of force in an armed conflict, provide essential guidance: only combatants are to be intentionally targeted (the principle of discrimination),

DOI: 10.4324/9781003312925-3

collateral damage is only acceptable to the extent that it is proportional to the value of the target being attacked (the principle of proportionality), and force may only be used if, and to the extent, it contributes to achieving a legitimate military end (the principle of necessity). These principles, however, only address ethical challenges involving the application of armed force. They also do not resolve many ethical quandaries which *do* involve the use of force – they do not, for example, offer any help to the patrol leader in the scenario described earlier.

What is needed, therefore, is a model or process for military ethical decision-making (EDM) which is both broader in scope than the principles of the *jus in bello* and which is at the same time consistent with them. That model must, furthermore, be both *practical* – an effective guide to ethically sound action in a wide range of situations – and *appropriate* as a tool for the military servants of liberal democracies. This latter consideration, the appropriateness of models of EDM in the context of the specific values and political foundations of liberal democracies, is largely overlooked but is nonetheless fundamental. It is with this issue, therefore, that I begin.

Appropriate Ethical Decision-Making by the Military Forces of Liberal Democratic Governments

Arguably, outside of the punitive frameworks of military justice, the most common means military forces have traditionally employed to direct soldiers'[1] EDM has been efforts to shape *character*. Put in philosophical terms this is, broadly speaking, a virtue approach to EDM. In virtue theory, it is excellences of character (virtues) that dictate right action, and right action is identified as that which a fully virtuous person would choose under relevantly similar circumstances. Stoicism, in particular, is a virtue theory that military personnel have long found attractive and that has been absorbed into military culture. Nancy Sherman goes as far as to identify Stoicism as 'the ancient philosophy behind the military mind' (Sherman 2007).

Despite the popularity of this virtue-based approach, in the context of a liberal democracy which seeks (within broad and accommodating boundaries) to maximize citizens' freedom to pursue their own conceptions of the good life, there are good reasons to consider the idea of the state deliberately seeking to shaping an individual's character, to be more than a little problematic. Consider, for example, this comment by Israeli military ethicist Asa Kasher:

A conception that constitutes a certain professional ethics is a conception of proper behaviour. An interesting alternative to behaviour and the principles that should guide it could be character and the virtues that are its parts. . . . We prefer principle-guided behaviour as the subject matter of military ethics in the context of [the Israeli Defence Force] . . . mainly because a military force of a democracy that includes people who are conscripts and people who are reserve officers and NCOs should educate them to follow the principles of military ethics necessary for the effective functioning of the military force, but avoid any attempt to change their character in a deep and broad way of long lasting effect. Respect for the human dignity of a conscript or a reserve officer means respect for their nature and liberty to the largest extent compatible with their ability and commitment to do their best in carrying out their military missions.

(Kasher 2008, 139–140)

While Kasher limits the application of his comments to conscripts and reservists, it seems to me that these considerations should also extend to full-time members of an all-volunteer force. Treating them differently is to treat them as in some fundamental way separate from society, as a kind of 'warrior class'. This is troubling in a number of ways, not least because this must inevitably exacerbate the civil–military gap, a constant source of problems in ensuring appropriate civil–military relations in a democratic society. So, if we are to resist this classification of the military, the issue Kasher raises seems to apply just as forcefully to full-time career military personnel.

Key to Kasher's point is the distinction he draws between principle-guided behaviour and behaviour that is directed through shaping the character and virtue of the individual in question. He writes,

People in uniform vary in their attitude to the very idea of an organizational conception of proper behaviour, towards the idea of a code of ethics in general or a military code of ethics in particular, or towards certain values or principles. Eventually, every man and woman in uniform should comply with the norms of their military ethics, on grounds of adequate understanding and sincere undertaking. The variety of preliminary attitudes must be taken into account when successful methods of presentation, explanation and persuasion are designed.

(Kasher 2008, 134)

Kasher's recognition of the normative weight of the diversity of moral perspectives held by uniformed personnel is important and largely overlooked in existing scholarship and practice. As citizens of a liberal democracy, the individuals who serve in the armed forces have the right to pursue their own conceptions of the good, so long as these conceptions fall within the broad bounds of liberalism (because, as Charles Taylor memorably puts it, 'liberalism can't and shouldn't claim complete cultural neutrality. Liberalism is also a fighting creed' (Taylor 1995, 249)).

It is easy to see why efforts to instil virtues in the citizens who are the armed servants of the liberal state are problematic when we remember that the dominant conceptions of virtue are *eudaemonist* – that is, they define the virtuous character in terms of some or other conception of 'human flourishing', a specific vision of what constitutes the good, or meaningful or valuable life. Such specificity is far too narrow to accord with the foundational impulse of liberalism. At the core of liberalism is what Gaus calls the Fundamental Liberty Principle, the view that 'freedom is normatively basic, and so the onus of justification is on those who would limit freedom, especially through coercive means' (Courtland *et al.* 2022). Related to this is the idea of liberal neutrality. Within the limits appropriately defined by the Fundamental Liberty Principle, the state must seek to allow as much space for individual freedom as possible and thus seek to avoid enacting laws and following policies that favour or disfavour any of the conceptions of the good life that comport with the bounds of liberalism. As Andrew Mason rightly states,

> The idea that respect for persons requires the state to be neutral (in some sense) between different conceptions of the good life is an important component of contemporary liberal theory. It is attractive at least partly because it resonates with the thought that the state should be impartial in relation to its citizens. Liberals themselves have varied accounts of what it is for the state to be neutral between conceptions of how to live, but they concur in thinking that neutrality has substantive political consequences.
>
> *(Mason 1990, 433)*

Though it is widely accepted that state neutrality regarding ways of life is an essential feature of liberalism, this acceptance is not universal.[2] Nonetheless, I take it that Peter Balint is correct in his analysis of the three aspects of liberal neutrality. According to Balint, liberal neutrality is generally conceived of in terms of neutrality of justification, neutrality of intent, and neutrality of outcome:

> Under neutrality of justification, no law of policy should be justified by the rightness of any particular way of life. This is a form of procedural neutrality, where the laws and policies that citizens should live by should be equally justifiable to all. . . . The second type of neutrality is of a different order. Here what matters is not so much how a law or principle has been justified, as much as its intention. Under neutrality of intent, a neutral institution or policy should not intend to favour (or hinder) any particular way of life. . . . The third type of neutrality is effectively the flipside of neutrality of intent. Neutrality of outcome is concerned not only that institutions of policies do not intend to favour any particular way of life, but that they do not actually favour – even unintentionally – any way of life.
>
> *(Balint 2015, 498–499)*

For our purposes, the key thing about state neutrality is that it defines the liberal state as one with a moral basis which we might describe as 'thin' and 'narrow' – based primarily on the commitments to liberty and security which provide the basis for the social contract. This is in strong contrast with the 'broad', 'deep', and 'rich' nature of individual character.

Still, we do recognize that society may place particular ethical demands on particular members of society because of the particular role they play within society.

This is the idea of 'role morality' or what Carl Ficarrotta calls 'the functional line': 'general demands on the character and behavior of military professionals mostly of the strictness variety that flow directly from the military function itself' (Ficarrotta 2010, 6). This corresponds with Kasher's minimalist focus on 'the principles of military ethics necessary for the effective functioning of the military force', and it seems to me that Kasher is correct that shaping character is not the goal, rather it is shaping behaviour that matters for this purpose.

This functional narrowing of the behaviour of military personnel does, obviously, have some impact on the range of possibilities available to them in living out their conception of the good life. However, this is less problematic given the critical point that in an all-volunteer force these individuals have (so long as they have been appropriately informed as to the consequences of volunteering for service) consented to be in this situation – this is in a real sense the result of them choosing a particular conception of the good or valuable or meaningful life. Furthermore, we should recollect Peter Balint's point about how to understand the neutrality of the liberal state. Neutrality can be maintained, even where choices are narrowed, if the neutrality of *justification* and neutrality of *intent* are maintained by the state, and there seems no reason to see this 'functional line' as being in violation of that.

These functional requirements do not, as Ficarrotta notes, amount to an expectation that military personnel be held to a 'higher' ethical standard. Ficarrotta encourages us to

> [notice] that the functional line might be applied to any enterprise, especially cooperative ones. To the degree that any undertaking is important, then we at once have special reasons for more strictly binding those engaged in the enterprise to general [ethical] standards that are necessary for its success.
>
> *(Ficarrotta 2010, 6)*

So, what is needed is that military personnel adhere to *particular* ethical requirements and do so *strictly* – the strict application of these requirements is demanded by the severity of the consequences of failure to adhere to them.

We must therefore approach with care the well-known statement by General Sir John Hackett that a morally bad person 'cannot be . . . a good soldier, sailor or airman' (Hackett 1986, 119). The thin and narrow morality of the liberal state does not allow the military arm of the state to define some conception of what the morally good life for the individual in uniformed service is. While functional requirements necessitate common adherence to particular norms within the military, and while the choice to serve in the military does narrow the scope of options for individuals regarding conceptions of the good life (as an obvious example, it is not compatible with a pacifist world view), care must be taken that the state does not undermine the very norms that its justification rests on.

All of this still leaves an important question to be answered: if it is not appropriate for individual soldiers' moral intuitions to be conformed through some kind of character-shaping exercise, how will it be possible to ensure any degree of consistency in EDM across the force? One part of the answer is straightforward – military personnel must act in adherence to the ethics of war as outlined in the *jus in bello* principles of the just war tradition. But things are more challenging when it comes to addressing ethical quandaries – such as the one with which I opened this chapter – that are not amenable to resolution through straightforward application of the principles of the just war tradition. How do we avoid the situation in which, when faced with such ethical challenges, soldiers have nothing to fall back on in making their decisions other than the intuitions and leanings generated by their personal moral frameworks, and no way of accounting for their decisions except in those terms?

As Marcus Schulzke correctly points out, 'Military ethics and just war are usually analysed from the perspective of one of the three dominant traditions in western moral philosophy: Aristotelian virtue ethics, Kant's deontological moral theory, and utilitarianism'. (Schulzke 2013, 96) I have already set aside as inappropriate the idea of choosing character-based approaches like Aristotelian (or Stoic) virtue ethics as the primary basis for military EDM. So, which of the alternatives should be selected, deontology or utilitarianism? The natural tendency is for individuals to choose the theory which most comfortably comports with their personal leanings and biases – but that will be different for different members of the military, so this approach amounts to personal preference. This is unsatisfactory, as military ethical decisions are first and foremost the decisions of *agents of the state*. And there seems no way to identify one of these approaches to military ethics as the sole appropriate guide for those applying military force on behalf of the liberal democratic state. As I discuss elsewhere (Baker 2020), following Charles Taylor, there are three 'perspectives' of the modern self that make up the DNA of the idea of the liberal democratic state. These are as follows: the Rationalist Perspective, with its deep emphasis on the rights and dignity of the individual; the Naturalist Perspective, with its commitment to the greater good for all; and the Inner Perspective, which radically individualizes the good. I argue further that from these emerge two fundamental drivers embedded within the social contract that defines the liberal state. First, there is the deep commitment to human rights which comes to us via the Rationalist Perspective, based on the *deontological* Kant-inspired notion of human dignity that we moderns take as axiomatic. Often obscured is the second: the fundamental rationale of entering into the social contract in the first place, which is to ensure the protection of those who fall within its scope. The latter consideration is *consequentialist*, or at the very least prudential, and lines up broadly with the Utilitarian position that is the most prominent representative of what I have called the Naturalist Perspective of modernity. These considerations exist in a virtuous tension that is foundational to liberal democracy – to remove one or the other is to fundamentally destabilize and distort the nature of our political system.

Accordingly, in order to do justice to both the rights imperative within liberalism and the consequentialist/prudential imperative for the state to protect its citizenry (the prime justification for the social contract), states must pursue policy in the international realm that reflects *both* of these fundamental commitments. The ethics of war for the liberal state must unquestionably take strong cognizance of the human rights of individuals affected by war, but the liberal state must at the same time stay true to its responsibility under the social contract to ensure the protection of its citizens. While I cannot pursue the issue here, this means that liberal states must be committed to a particular interpretation of the ethics of war, one which rejects the exclusive focus on personal rights characteristic of revisionist approaches to the just war tradition such as that of Jeff McMahan, on the one hand, but one which also rejects unprincipled realism, on the other. Happily, this is a position that rests comfortably within the just war tradition, even though that tradition makes no assumptions about the nature of the political foundation of the state.

The implication of this is that to be appropriate – that is, in alignment with the DNA of the foundational ideas of liberal democracy – whatever model of EDM is adopted by the military of such a state must account for both the deontological and the consequentialist foundations of the liberal democratic state. Furthermore, some acknowledgement of the intensely personal Internal perspective – which loosely correlates to virtue ethics – is also appropriate, even if it cannot be the primary driver. In what follows, I expound a model of EDM that meets this requirement and that is at the same time practical: the Ethical Triangulation model.

Practical Ethical Decision-Making: The Ethical Triangulation Model

As we have seen, Marcus Schulzke points out that 'Military ethics and just war are usually analysed from the perspective of one of the three dominant traditions in western moral philosophy: Aristotelian virtue ethics, Kant's deontological moral theory, and utilitarianism' (Schulzke 2013, 96). Furthermore, he notes, there is a significant problem with this plurality of perspective: 'Each of these theories offers its own decision procedures and reasons for acting, and *each may lead to radically different conclusions about what is the right course of action*' (Schulzke 2013, 96, italics added). Clearly, even setting aside questions of appropriateness (as discussed in the previous section), this is a practically undesirable outcome for a military force. US Special Forces Major Peter Dillon was certainly being overly reductionist when he wrote that 'the goal of battlefield ethics is unsupervised predictability of soldier conduct' (Dillon 1992, p iii, note 12), but it is undoubtedly true that consistency of behaviour is a highly desirable outcome of ethics education in the military.

Each of the dominant approaches mentioned by Schulzke has its own appeal. Indeed, for most of us, each of these approaches resonates deeply. The idea that consequences matter in ethics has strong intuitive appeal, as does the notion that others are deserving of our respect and should not be treated merely as a means to an end. Likewise, there are few of us who don't feel the pull of character-based concepts like courage, integrity, compassion and honour. However, despite the appeal of each of these approaches, each of them also has its own challenges.

Consider first Utilitarianism, the dominant form of consequentialism. This approach seems to have trouble accounting for the importance of the individual and sometimes seems to give rise to counterintuitive results. From a specifically military perspective, Lieutenant Colonel David Glendenning points out,

> Divorcing outcomes from motives, although morally logical, can lead to ethical cul-de-sacs and may not offer firm ground when trying to appease your conscience after an event. Similarly, the unpredictable nature and fast tempo of military events removes the epistemic certainty to accurately predict the second and third order consequences of any given action. Another danger of this approach is slipping towards ethical egoism whereby one, consciously or subconsciously, prioritizes personal utility which could be at odds with the selfless nature of military service. Lastly, a strictly utilitarian reasoning model creates an analytical reliance that could constrain a flash of military brilliance and intuition within a military planning cycle. In this sense, operational art and consequentialism are not necessarily mutually reinforcing.
>
> *(Glendenning 2017, 27)*

Deontological (or rule-based) theories of ethics, such as that developed by Immanuel Kant, are also problematic in some regards. They can sometimes seem too rigid and blind to the importance of consequences in certain circumstances.

> The common problems associated with a rules-based approach to ethics include a failure to properly consider consequences and, at times, the production of a counterintuitive result where rules may inhibit a rational response. . . . [T]rying to overlay a set of rules to an ever changing operational context is laced with complexity.
>
> *(Glendenning 2017, 28)*

The third main approach to ethics is virtue ethics. With its emphasis on 'achieving what is good rather than avoiding what is bad', and its focus on character, this is both an appealing and a common approach in military training (Glendenning 2017, 29). But virtue approaches to ethics, with their focus on what we should *be*, suffer from a significant drawback: they don't seem to offer us enough clear guidance about what we should do when dealing with moral dilemmas.

This is, of course, by no means a comprehensive account of the challenges faced by each of these approaches to ethics, or of the many sophisticated ways in which adherents to each approach have sought to address these issues. This is what we might call the *philosophical* approach to ethics. For our purposes, however, a more comprehensive account is not needed – the goal here is not to choose one approach over the others but instead to articulate a mechanism that enables the best features of each to come to the fore and which at the same time compensates somewhat for the shortcomings of each. Put differently, what I am advocating here is what we might call the *pragmatic* approach. This is, in practice, what many well-trained applied ethicists do – apply each of the different approaches and concepts as they best fit the circumstances concerned.

What makes this *pragmatic approach* work for professional ethicists is their deep and intuitive understanding of the theory. For most non-specialists, however, there is value in having an explicit methodology to guide us. For military personnel, the value is more specific. Having a methodology which (at least broadly) maps to the three perspectives of modernity which so shape the societies of liberal democratic states, and which also provides a common language for ethics, provides a more appropriate alternative to simply falling back on the intuitions and leanings of individual moral frameworks when ethical dilemmas must be faced. This section outlines one such approach, one I have employed over years of teaching ethics to military officers in training, first at the US Naval Academy and now at the Australian Defence Force Academy.

I call this approach Ethical Triangulation. The idea of 'triangulation' resonates with military personnel because of the importance of navigation in the military, and the idea fits well with the tripartite nature of mainstream ethical theory. In its simplest form, the idea is to, as it were, 'take a bearing' from each of the main approaches to ethics when considering an ethically challenging question. This ensures that the decision maker does not overlook any important ethical considerations that might be obscured by applying only one ethical approach, and also benefits from the fact that the strengths of each of these approaches often fall in exactly the ethical 'space' that the weaknesses of one or both of the other approaches are in.

As an explicit methodology, Ethical Triangulation works like this. The first step is to take a bearing from the ethical peak of 'Respect' (deontology), by asking what ethical principles or rules apply to the situation we are considering. For some, this is a little counterintuitive as, in my experience, many want to begin by examining potential consequences. But there are good reasons for starting with deontology rather than consequentialism. For one thing, deontological theories of ethics represent what we might think of as the collected ethical wisdom of the ages, and the principles embedded in deontological ethics are often (as rule Utilitarians recognize) a good guide to achieving the best consequences. Perhaps more importantly, starting here represents a recognition of the first imperative (as so ably identified by former British Army 'ethics Tzar' Phillip McCormack (2015)): to respect the rights and dignity of the individual. (While deontological principles are many and varied, they do, as a broad rule of thumb, reflect this imperative.) Add that to one of the key limitations of consequentialism – that we're often poor at predicting consequences, particularly when self-interest is involved – and we have good reason to start with deontology.

Once the decision maker has identified the deontological principles that apply to the case in question (which I refer to in my teaching to military personnel as 'Constraints and Commitments'), they will either have a first suggestion of what the ethically appropriate action or actions should be or it will become clear that there are competing principles which suggest different courses of action. The next step is to look to the horizon again and consider the possible consequences of the course or courses of action which we are weighing up. This is the 'Consequences' peak. In so doing, the decision maker might find that the potential consequences of the course of action suggested by his or her assessment of the deontological principles are serious enough, and certain enough, that those consequences outweigh the deontological principles. This will then point us to a different ethical solution. Alternatively, it may be that weighing up the consequences favours one course of action over the alternatives in those cases where conflicting deontological principles suggest different ways forward.

The third step is to take a final bearing from the ethical peak of 'Character' (which connects us, loosely, to virtue ethics). This step shifts the decision maker's attention, for a moment, away from the issue under consideration onto himself or herself. It is a moment of self-reflection in which the decision maker must weigh the impact her moral identity may be having on her assessment of the ethical problem and its potential solution. The goal is to recognize and, if necessary, compensate for known biases and blind spots. This will, of course, be meaningless if the decision maker does not know her character well, and here, it is worth reminding ourselves of one of the most famous quotes in all of philosophy, alleged to have been uttered by Socrates, that 'the unexamined life is not worth living' (Plato 1966, 38a). Just as a strong magnet can throw off the needle of a compass, the decision maker's moral identity can, if she is not careful, affect her ethical decision-making and potentially point her in an ethically problematic direction. It is important to note that including this self-reflective step in the Ethical Triangulation decision-making model does not constitute character shaping. It is a genuinely individual assessment and as such reflects well the Inner Perspective of modernity. It can potentially also bring clarity to the earlier steps in this process of Ethical Triangulation.

Deliberately working through these three steps of Ethical Triangulation should generally help military personnel to make better ethical judgements than they would otherwise. In many cases – perhaps even most – that will be all that is needed. Ethics is not, however, a perfect science, and while this process will certainly help us to exclude a significant number of possible courses of action as clearly inappropriate, it is possible that sometimes there will still be an unresolved conflict between two or more possible options that could perhaps be the right thing to do. In this grey area in which theory-driven analysis is sometimes inadequate to the task of giving final guidance, it becomes a matter of wisdom of how to proceed. This reflects, though I think in a non-problematic way, the virtue ethics idea of *phronesis* or practical wisdom. Practical wisdom is the ability to appropriately marry up right intentions and sound character with practical action. It is the product of experience and is often best developed through learning from those wiser than ourselves. Some will find this lack of analytic precision in such cases to be frustrating, but we would do well to remember the wisdom of Aristotle himself, who reminds us that

> it is the mark of an educated [person] to look for precision in each class of things just so far as the nature of the subject admits; it is evidently equally foolish to accept probable reasoning from a mathematician and to demand from a rhetorician scientific proofs.
>
> *(Aristotle 1984, Book I, 1094.b24)*

This, then, is the Ethical Triangulation model as I teach it. As mentioned earlier, it is an attempt to make explicit a heuristic that reflects how applied ethicists tend, in practice, to engage with ethical challenges. As such, it should come as no surprise that the idea of approaching ethical challenges from these three perspectives is widespread in applied ethics (as opposed to academic philosophical ethics) in various fields. For example, as far back as the 1990s James Svara, Research Professor in the School of Public Affairs at Arizona State University and Visiting Professor in the School of Government at the University of North Carolina, Chapel Hill, was expounding a triangulation model in the context of ethics in Public Administration, and included this in his influential primer on ethics in public administration (Svara 2015). The Brown University Science and Technology Studies (STS) Program explains that there are three main types of ethical theory – consequentialist, non-consequentialist, and agent-centred – and contends that from this that 'it makes sense to suggest three broad frameworks to guide ethical decision making: The Consequentialist Framework; The Duty Framework; and the Virtue Framework'. As the nine contributors to the Brown University *STS Framework for Making Ethical Decisions* explain, each of these 'is useful for making ethical decisions, none is perfect – otherwise the perfect theory would have driven the other imperfect theories from the field long ago'. Consequently, for applied ethical decision-making they advocate 'putting the frameworks together'.[3]

Given the broad currency of the model, it is unsurprising that it has found traction in military circles. My own initial published account of Ethical Triangulation (Baker 2015) was followed within a year by a paper by US Command and General Staff College Professor, Colonel (Ret.) Jack D. Kem, PhD, entitled 'Ethical Decision Making: Using the "Ethical Triangle"' (Kem 2016). Kem does not refer to my account and appears to have come to the approach independently thereof. Kem begins with a discussion of the Army's leadership manual (ADRP 6–22) which he contends 'does not provide a lot of guidance on how to address ethical issues, other than to embrace the Army Values' – a proposal he finds inadequate to the task of decision-making in the face of challenging ethical dilemmas. Instead, he affirms the proposal that US Army leaders should consider ethical dilemmas from three perspectives:

> Leaders use multiple perspectives to think about ethical concerns, applying the following perspectives to determine the most ethical choice. One perspective comes from the view that desirable virtues such as courage, justice, and benevolence define ethical outcomes. A second perspective comes from the set of agreed-upon values or rules, such as the Army Values or Constitutional rights. A third perspective bases the consequences of the decision on whatever produces the greatest good for the greatest number as most favorable.
>
> *(ADRP 6–22, para 3–38 (2012 edition))*

Taking his lead from this, Kem groups ethical theories into three main groups: principles-based ethics, consequences-based ethics, and virtues-based ethics. He proposes that the best approach for practical EDM is to apply the 'ethical triangle'. As he notes:

> There are a number of questions that could be asked about these three perspectives: Which of the ethical philosophies are the most useful – rules or principles-based ethics, utilitarian or consequences-based ethics, or virtues-based ethics? Which one of the philosophies is the best fit for human behavior? All three appear to have some merit; all three can be used for decision-making as "distinct filters that reveal different aspects of a situation requiring an ethical choice." To only consider one of the different theoretical bases runs the risk of being one-sided in analysis. Whether principles, consequences, or virtues provide the true reasons for ethical decision-making, all three of the theories and their lineage are useful for gaining insight into the complexity of ethical decision making.
>
> *(Kem 2016, 4–5)*

While Kem does not emphasize this, his model is consistent with my account of Ethical Triangulation with regard to the order in which each perspective is considered: principles-based ethics first, then Consequences-based ethics, and finally virtues-based ethics.

According to a recent article, Kem's model is taught at the US Army Command and General Staff College, the US Naval Postgraduate School, and the US Joint Special Operations University (Ordiway 2022, 9). There is also evidence that there is at least awareness of this model among instructors at the US Army NCO Leadership Centre of Excellence (Rivera 2018), and in the US military medical community (Martinez and Martinez 2017). My own Ethical Triangulation model has also achieved currency in military ethics education outside of the United States. For example, Professor David Whetham (Director of the Kings College London Centre for Military Ethics) teaches it at the UK Command and General Staff College and other UK military education and training institutions. The KCL Centre for Military Ethics online curriculum, the foundation of the Colombian Armed Forces military ethics programme and studied worldwide, incorporates the model.

Ethical Triangulation and the Just War Tradition

While it is common to hear of the 'just war theory', this is an inaccurate description. Instead, the accepted principles that guide the ethics of armed conflict have emerged through a millennium-long process of negotiation and compromise, hence the more accurate description, the 'just war tradition'.[4] This tradition reflects no single theory of ethics but is instead a pragmatic merger of a range of ethical considerations. The just war tradition is similar to the Ethical Triangulation method – both are ethical decision-making methodologies rather than ethical theories. There is a remarkable degree of compatibility between the Ethical Triangulation method and the principles of the just war tradition.

Consider, first, the principles of the *jus ad bellum*, which establish the conditions under which it is appropriate for a state to choose to go to war. The Ethical Triangulation method begins with identifying the 'Constraints', or deontological rules, that apply. In the case of the *jus ad bellum*, there are two critical deontological considerations: Just Cause and Legitimate Authority. A party must have a Just Cause (usually self-defence) and must be a Legitimate Authority (usually a legal head of state) to have a right to wage war. The Ethical Triangulation method then asks the decision-maker to consider potential 'Consequences' and weigh those in the light of deontological considerations. In the *jus ad bellum*, it is the principles of proportionality, the likelihood of success, and last resort that require the decision-maker to weigh the foreseeable outcomes of waging war against the right to do so. The final step in the Ethical Triangulation method is to consider 'Character' or virtue. Likewise, the *jus ad bellum* requires the decision-maker to evaluate their intentions, demanding that only the intention of achieving the just cause should drive the decision to go to war.

Or consider the *jus in bello*. The Ethical Triangulation method begins with identifying the 'Constraints and Commitments', or deontological rules, that apply. In the case of the *jus in bello*, the deontological principle of discrimination is the bedrock that provides both the central permission and constraint for the combatant – they are permitted to use lethal force against combatants but may not intentionally target non-combatants. The Ethical Triangulation method then asks the decision-maker to consider potential 'Consequences' and weigh those in the light of deontological considerations. In the *jus in bello*, the principles of proportionality and necessity fill this role. The principle of proportionality requires the decision-maker to evaluate whether the foreseen (but not intended) collateral consequences of their proposed course of action (harm to non-combatant persons and property) might outweigh the deontological permission to engage enemy combatants in that particular set of circumstances. In a similar vein, the principle of necessity requires the decision-maker to evaluate, for any act of lethal targeting, whether the deontological permission to engage enemy combatants is undermined by a lack of sufficient military advantage accruing from the foreseeable harms to combatants likely to result from that strike. The final step in Ethical Triangulation is to consider 'Character' or virtue. This aspect relates to the *jus in bello* principle of humanity, which forbids the use of methods and means of war considered inhumane because they cause unnecessary suffering.

A Critique of Ethical Triangulation

In a recent article Benjamin Ordiway, a US Army Civil Affairs Officer and now an instructor at the US Military Academy (West Point), has offered a critique of Kem's 'ethical triangle' approach to military decision-making. As we have seen, Kem's approach corresponds closely with my account of ethical triangulation. Ordiway is not entirely dismissive of this approach, noting that 'the Ethical Triangle offers a practical way to frame a problem and exposes the decision-maker to normative frameworks (i.e. deontological, consequentialist and virtue approaches)' (Ordiway 2022, 9). Nonetheless, he contends that this approach is fundamentally flawed[5]:

> The main problem with the Ethical Triangle is that nobody, when personally involved in a situation with significant moral content, thinks like this. When faced with a situation – say, one where your character, career and reputation are on the line – a typical first step is to experience emotional turmoil. A next possible step is to frame the situation as a threat or benefit to self and employ moral disengagement mechanisms or to rationalize a way out. In short, in dispensing with system one considerations, the Ethical Triangle neglects the very humanity of the humans it is supposed to guide. Emotions will color one's interpretation of the situation and any duties or obligations connected to it for better or worse. It would be best, then, to confront our emotions and extract the useful moral information they may contain while being alert to their tendency to bias us in unhelpful ways. By grooming SOF for a rational, system two approach as typified by the Ethical Triangle, PME faculty miss the opportunity to educate students on how to retain the moral content of their emotions.
>
> *(Ordiway 2022, 9)*

Ordiway's comment about 'system one' is a reference to the influential work of Nobel prize-winning economist, Daniel Kahneman. As Ordiway explains:

> In his 2011 book Thinking, Fast and Slow, Daniel Kahneman popularized the concept of two unique yet overlapping cognitive processes: "system one" and "system two." Beginning with the latter, system two thinking is deliberate, slow (relative to system one), and effortful. We tend to think of ourselves as supremely

rational, utility-maximizing decision-makers. We are, but only rarely. This rarity is because system two thinking is mentally and physically taxing. Just imagine slogging through life with each decision requiring the focus you might (should!) apply to a change of command inventory – every day would be an all-nighter. Unlike system two, system one thinking is intuitive, rapid, involuntary, associative and requires minimal effort. It often employs mental shortcuts which are unquestionably helpful and accurate in many circumstances. For example, if you have ever driven home from work without recalling the journey, system one was at the wheel. Still, system one can lead us astray.

(Ordiway 2022, 7)

What should we make of Ordiway's critique? We should, first, wholeheartedly agree with Ordiway as to the importance of system one cognitive processes in determining our behaviour. Earlier attempts to get to grips with this invoked the distinction between 'tests of integrity' and 'ethical dilemmas' – in the former, the right thing to do is clear, it's just hard to do it, and in the latter, it is difficult to determine the ethical course of action. This distinction, while helpful, obscures the fact that it is by no means only a lack of integrity that leads people to act in ways that they would otherwise acknowledge to be unethical. The work of Kahneman and others like Robert Sapolsky (Sapolsky 2017) has shown us just how easily our system one processes are influenced by biological and situational factors. In my teaching of military ethics to military personnel, and related writing,[6] I draw heavily on the work of Deanna Messervey who has done us the great service of drawing together a wide range of relevant research in moral psychology and showing its relevance to military decision-making.[7]

Ordiway is, however, mistaken in contending that recognizing the role of system one cognitive processes in ethical decision-making entails that something like Kem's Ethical Triangle or my Ethical Triangulation is not useful. Kahneman's work helps us to see that we are *both* system one *and* system two thinkers. Ordiway is right that we must not overlook system one cognitive processes, but it would be just as much an error to overlook system two cognitive processes. There are many circumstances in which ethical decisions can, and should, be made on the basis of clear, deliberate, and unemotional evaluation. Furthermore, noting that (as Ordiway acknowledges) system one can 'lead us astray', there is good reason to encourage military personnel faced with challenging ethical decisions to engage system two whenever possible, and applying Ethical Triangulation is a mechanism by which to do that. As Roger Herbert, retired SEAL and co-author of this volume, is fond of saying to military personnel he is teaching and training, 'you often have more time than you think – stop the bleeding, smoke a cigarette,[8] and think it through'. There is another aspect to this. There is some early indication from an internal study carried out in the Australian Defence Force that training in ethical triangulation can serve to reduce cognitive load in military personnel facing challenging ethical decisions in high-pressure environments.[9] This possible effect has yet to be established through comprehensive and rigorous experimental research. However, if it turns out to be a sound hypothesis, this would constitute empirical evidence that Ethical Triangulation, though a system two 'tool', can impact positively EDM in situations in which system one cognitive processes naturally dominate.

Conclusion

Lieutenant Colonel Glendenning comments on the potential (in the context of the British Army) of adding Ethical Triangulation to the conceptual tools imparted to military personnel:

> Teaching ethical leadership and ethical theory triangulation will provide a functional theoretical framework to structure the delivery of ethics education in the British Army. . . . Recognising the various ethical lenses that will interpret dilemmas from a different perspective offers a powerful frame of reference to design an ethical educational pathway for the British Army . . . a necessary stepping-stone towards creating moral autonomy.
>
> *(Glendenning 2017, 30)*

Ethical Triangulation is not a theory of ethics. The methodology does not aim to reconcile the very different (and, arguably, irreconcilable) philosophical ideas underpinning the deontological, consequentialist, and virtue-based approaches to ethics that are in a 'virtuous tension' at the foundation of our liberal democratic tradition. Instead, it is a *practical decision-making methodology* designed to take cognizance of the core ethical intuitions inherent in each of these theoretical approaches as a way to avoid overlooking key considerations that might be relevant to the ethical decision in question.

Notes

1 Here, and throughout this chapter, I will use 'soldier' as shorthand for any uniformed member of the military, including uniformed members of naval, marine, air force, and other military formations.
2 I discuss this issue in more detail in Deane-Peter Baker, *Morality and Ethics at War: Bridging the Gaps Between the Soldier and the State* (Bloomsbury Academic, 2020), 45–47.
3 www.brown.edu/academics/science-and-technology-studies/framework-making-ethical-decisions
4 See, for example, Morkevicius (2018).
5 Ordiway offers another substantive critique of Kem's Ethical Triangle model; however, as this relates to a feature that is not intrinsic to ethical triangulation, I have not addressed that here.
6 See, for example, Baker (2020), especially Chapter 6.
7 See, for example, Messervey and Peach (2014).
8 Figuratively, of course. Herbert is not a smoker nor an advocate thereof!
9 This research was carried out by Warrant Officer Matthew Wann, Australian Army.

References

Aristotle. 1984. 'Nichomachean Ethics' in Jonathan Barnes (ed.) *The Complete Works of Aristotle: The Revised Oxford Translation*. Princeton University Press, 1729–1867.

Baker, Deane-Peter. 2015. 'Ethical Triangulation' in Deane-Peter Baker (ed.) *Key Concepts in Military Ethics*. UNSW Press.

Baker, Deane-Peter. 2020. *Morality and Ethics at War: Bridging the Gaps Between the Soldier and the State*. Bloomsbury Academic.

Balint, Peter. 2015. 'Identity Claims: Why Liberal Neutrality Is the Solution, Not the Problem.' *Political Studies* 63(2), 495–509.

Courtland, Shane D., Gerald Gaus, and David Schmidtz. 2022. 'Liberalism' in Edward N. Zalta (ed.) *The Stanford Encyclopedia of Philosophy*. Metaphysics Research Lab, Stanford University.

Dillon, Peter J. 1992. 'Ethical Decision Making on the Battlefield: An Analysis of Training for U.S. Army Special Forces.' Master of Military Art and Science Thesis, US Army Command and General Staff College.

Ficarrotta, J. Carl. 2010. *Kantian Thinking About Military Ethics*. Routledge.

Glendenning, David. 2017. *Who Really Sets the Bearing on My Moral Compass? An Assessment of the Utility of Moral Autonomy in the Contemporary Operating Environment*. UK Joint Services Command and Staff College/EuroISME.

Hackett, John Winthrop. 1986. 'The Military in the Service of the State' in Malham M. Wakin (ed.) *War, Morality and the Military Profession*. Westview Press, 104–119.

Kasher, Asa. 2008. 'Teaching and Training Military Ethics: An Israeli Experience' in Paul Robinson, Nigel de Lee, and Paul Carrick (eds) *Ethics Education in the Military*. Routledge, 133–146.

Kem, Jack D. 2016. 'Ethical Decision Making: Using the Ethical Triangle' (Command and General Staff College Foundation). www.cgscfoundation.org/wp-content/uploads/2016/04/Kem-UseoftheEthicalTriangle.pdf (accessed 04/05/2023).

Martinez, Katie, and Marcos Martinez. 2017. 'Application of the Ethical Triangle in the 2014 Ebola Epidemic: A Case Study.' *InterAgency Journal* 8(4), 34–46.

Mason, Andrew. 1990. 'Autonomy, Liberalism and State Neutrality.' *Philosophical Quarterly* 40(160), 433–452.

McCormack, Philip J. 2015. *Grounding British Army Values Upon an Ethical Good*. Command and General Staff College Foundation.

Messervey, Deanna L., and Jennifer M. Peach. 2014. 'Battlefield Ethics: What Influences Ethical Behaviour on Operations?' in Gary Ivey, Kerry Sudom, Waylon H. Dean, and Maxime Trembley (eds) *The Human Dimensions of Operations: Personnel Research Perspective*. Canadian Defence Academy Press, 83–101.

Morkevicius, Valerie. 2018. *Realist Ethics: Just War Traditions as Power Politics*. Cambridge University Press.

Ordiway, Benjamin. 2022. 'Developing SOF Moral Reasoning: Preparing Humans for Hard Wear on the Moral Terrain.' *Special Warfare*, 6–11.

Plato. 1966. *Plato in Twelve Volumes*, Vol. 1, trans. Harold North Fowler. Harvard University Press.

Rivera, Jorge A. 2018. 'Using an Ethical Framework for a More Responsible Online Image'. NCO Journal, May 18. www.armyupress.army.mil/journals/nco-journal/archives/2018/May/Ethical-Online-Image/ (accessed 04/05/2023).

Sapolsky, Robert. 2017. *Behave: The Biology of Humans at Our Best and Worst*. Penguin Press.

Schulzke, Marcus. 2013. 'Ethically Insoluble Dilemmas in War.' *Journal of Military Ethics* 12(2), 95–110.

Sherman, Nancy. 2007. *Stoic Warriors: The Ancient Philosophy Behind the Military Mind.* Oxford University Press.

Svara, James. 2015. *The Ethics Primer for Public Administrators in Government and Non-profit Organizations.* Jones and Bartlett Learning.

Taylor, Charles. 1995. *Philosophical Arguments.* Harvard University Press.

3

THE MORAL DELIBERATION ROADMAP

The US Naval Academy's Moral Reasoning Framework

Roger Herbert

In 2016, not quite five years after his US Naval Academy graduation, the officer-in-charge of Riverine Command Boat (RCB) 802 (a Navy Lieutenant, henceforth 'the OIC') confronted an ethical challenge as consequential as anything I faced in my 26-year career as a US naval officer.[1] In this chapter, I summarize the incident, consider how the Navy had prepared the OIC for that moment, and suggest some shortcomings in that preparation. I then introduce the Moral Deliberation Roadmap, the moral reasoning framework adopted by the US Naval Academy in 2021, about a decade *after* the OIC's graduation. Finally, I apply this framework to the incident to illustrate its strengths and limitations.

Incident at Farsi Island[2]

On 11 January 2016, Coastal Riverine Squadron-3 ordered RCB 802 and RCB 805 to depart Kuwait Naval Base and transit 250 miles to Naval Support Activity Bahrain for follow-on tasking. As the Commander, Destroyer Squadron Fifty (COMDESRON 50) investigation makes clear, the OIC's chain of command had not set him up for success. The crews, already stressed from a high-tempo deployment, were given only 24 hours to prepare for a transit that would be twice the distance either of the crews had ever attempted and require refuelling at sea, an inherently hazardous procedure. Threading the needle between Saudi and Iranian territorial seas would challenge the crews' navigation skills.

Further complicating matters was a convoluted chain of command. The OIC, the senior (and only) commissioned officer in the patrol, was Boat Captain for RCB 802, responsible for directing RCB 802 and its crew. He was not, however, designated as Patrol Leader, the person responsible for the overall execution of the mission. That responsibility fell to RCB 805's Boat Captain, a first-class petty

DOI: 10.4324/9781003312925-4

officer (a relatively junior non-commissioned officer or NCO).[3] Yet regardless of the decision to designate Patrol Leader duties to an NCO, the Navy, by regulation and custom, holds the Senior Officer Present Afloat, the SOPA, accountable for operations at sea. The OIC, in other words, was not responsible for the success and safety of the patrol, but he would be held accountable.

Frustrations continued into the morning of 12 January. After pulling an all-nighter, the crew of RCB 802 had successfully repaired an engine casualty. But just as their engine came back on line, the crew encountered satellite communication issues, delaying their planned departure by nearly four hours. The likelihood that they would make it to their refuelling rendezvous point before sunset, the OIC's strong preference, became even more remote when they were advised that the rendezvous location had moved 20 miles to the south.

The OIC decided to hedge his bet by revising the navigation plan. Although the OIC's shortcut improved their chances of reaching the at-sea refuelling point before sunset, the revised PIM (Plan of Intended Movement) failed to account for Saudi or Iranian territorial seas. This oversight would prove fateful.

Territorial seas typically extend 12 nautical miles from a state's territory. So, at 3:46 pm, when the OIC observed a landmass on the horizon, he knew they were no longer in international waters. Moments after this discovery, RCB 802's starboard engine started losing oil pressure, and the engineer shut down both engines to make repairs. RCB 805 came alongside. At 4:13, both boats were dead in the water, 1.6 nautical miles off the coast of Farsi Island, the yet-to-be-identified landmass.

Minutes after RCB 802's engine casualty, the crews observed two small boats speeding toward them. At first, the OIC suspected no hostile intent. Customary maritime law obliges vessels to assist other craft in distress. But he soon revised his estimation of the approaching vessels' intent. At about 300 meters, he observed the crewmembers of the approaching boats, each armed with an AK47, remove the covers of the deck-mounted weapons and jack the charging handles. The OIC recognized the streaming blue flags as the colours of the Islamic Revolutionary Guard Corps (IRGC).

After attempting unsuccessfully to communicate with the Iranians via marine band radio, both Boat Captains ordered their crews to 'kit up', man their weapons, and prepare them for action. Meanwhile, on RCB 802, the OIC urged his engineer to make haste while holding up a wrench and pointing to the engine compartment to signal the Iranians that they were not looking for a fight. At 4:28, with the Iranian boats rapidly closing, RCB 802's engineer shouted, 'We're good'!

The OIC's order, 'Let's get the fuck out of here. Go, go, go!' went out over the radio to RCB 805 and directly into the ear of RCB 802's coxswain. RCB 805 came up on step. But before RCB 802 could accelerate, the two Iranian boats, correctly assessing that RCB 802 was the less capable boat, manoeuvred in front and abeam of them. With fingers on their triggers, the Iranians took aim at the US sailors. RCB 802's coxswain, who the OIC describes as one of the best boat handlers in the Navy, could not manoeuvre around the Iranian boats. After extraordinary efforts, he dropped the engines into neutral and turned to his OIC with a frustrated, 'what do you want me to do?' expression on his face. 'Sir?'

So, what do you do now, Lieutenant? IRGC soldiers are preparing to board your vessel forcibly. Flight is no longer an option. Do you engage Iranians in a point-blank gunfight, or do you surrender your boat and crew?

The Education of an OIC

The Farsi Island incident offers a rich case study for USNA's Department of Leadership, Ethics, and Law (LEL). Not only does it stress each of the disciplines contained in the department's name, but it also features a highly relatable protagonist for our students. A few short years before his encounter at Farsi Island, the OIC was a Naval Academy Midshipman. He lived in Bancroft Hall, complained (groundlessly) about King Hall food, and studied military ethics in the classrooms of Luce Hall, overlooking the Severn River. Today, USNA students revisit the Farsi Island case during each of their core leadership courses.[4]

In this section, I consider how the Navy prepared – and, arguably, underprepared – the OIC for this legally and ethically complex decision. First, I outline the laws and regulations that informed his choices. I then address the OIC's formal education in moral reasoning.

The Legal and Regulatory Framework

US military vessels generally plan their routes to avoid foreign territorial seas when practicable. The OIC's revised PIM, which passed through both Saudi and Iranian territorial seas, was a breach of SOP. Importantly, however, it was *not* a violation of international law. The UN Convention on the Law of the Sea (UNCLOS) recognizes the 'right of innocent passage' for vessels, including warships, transiting through territorial seas. Unless the Americans had engaged in a 'non-innocent' activity as specified by the Convention (which they had not), they retained their right to non-interference during their transit.[5]

The IRGC, by contrast, committed at least two flagrant violations of international law. First, by cutting off RCB 802's egress, the IRGC vessels had 'prevented RCB 802 from exercising its right of innocent passage' (COMDESRON 50 2016, 153). Additionally, by attempting to forcibly board, the IRGC violated RCB 802's 'sovereign immunity'. According to UNCLOS, any military or other 'state-owned' vessel is subject to the jurisdiction *only* of the state whose flag they fly. In other words, they are 'immune from arrest or search, even when operating within the territorial seas of another coastal state' (Ibid., 154).

What courses of action (COAs) were legally available to the OIC? Given that Iranians had displayed 'capability, opportunity, and intent to commit . . . a hostile act' that an assault was imminent, and that escape was impossible, the OIC had the right to use proportionate force 'to defend his unit and his sovereign immune vessel'

(Ibid., 154). Furthermore, he was under no obligation to 'absorb the first round' (Ibid., 157, 103). The investigation confirms the RCB crews received recent training in the Fifth Fleet Standing Rules of Engagement (SROE) and the parameters of unit self-defence, so it's reasonable to assume that the OIC understood he had the legal authority to fight, to use lethal force to repel the boarders (Ibid., 101–104).

What is less clear is whether the OIC had a legal *obligation* to fight. According to the SROE, 'Unit Commanders always retain the inherent right and *obligation* to exercise self-defence in response to a hostile act or demonstrated hostile intent' (cited in Ibid., 100; emphasis added). Article II of the Code of Conduct (detailed in Executive Order 10631) is even more explicit: 'If in command, I will never surrender my men while they still have the means to resist'. The Code of Conduct recognizes that service members may be 'captured' after evasion and resistance become impossible, but they may never wilfully surrender if they can still fight (Ibid., 136). The penalties for doing so are stiff. According to the American Uniform Code of Military Justice (UCMJ; 10 USC 899, art. 99), 'Any member of the armed forces who before or in the presence of the enemy . . . shamefully abandons, surrenders, or delivers up any command, unit, place, or military property which it is his duty to defend' may be punished by death.

After the arrival of IRGC reinforcements – about 10 minutes after the initial encounter – the Americans were so badly outgunned that it's fair to say they had no means to resist. Prior to that, however, the US and Iranian vessels were more or less evenly matched. Given that both boats, especially RCB 805, were still capable of mounting meaningful resistance, did the OIC fail to carry out a legal duty to resist when he ordered his crew to stand down and 'do what they tell us' (Ibid., 108)?

The UCMJ and Code of Conduct seem unequivocal, but there is a critical complication. Relations between the United States and Iran were certainly unfriendly in 2016, even adversarial. But the two states were not at war. The OIC was not 'in the presence of the enemy'. Significantly, Geneva Convention protections afforded prisoners of war (Article 3) would not be in force. Had the Americans been captured after mounting a forceful – and foreseeably lethal – resistance, they would likely have been detained and tried as criminals, not POWs.

Members of the US Armed Forces are always bound by the Code of Conduct, in peacetime and in times of war. But the COMDESRON 50 investigating officer correctly observes that 'It is unclear how the Code of Conduct and related guidance . . . on "surrender" applies to non-armed conflict scenarios' (Ibid., 159). The Department of Defense instruction that guides Code of Conduct training (DoD Instruction 1300.21, 8 January 2001) acknowledges the tension between peacetime and wartime compliance with Article II.

[S]ervicemembers must be prepared to assess the dangers associated with being taken into captivity by local authorities. Their assessment of the dangers should dictate what efforts should be taken and what measures of force may be required to avoid capture, resist apprehension, and resist cooperation once captured.

Accordingly, the COMDESRON 50 investigating officer concludes, 'For operations other than war, my opinion is that . . . the DoD Instruction lacks any guidance on the issue of "surrender" and . . . does not apply to the RCB Boat Captain's surrender' (COMDESRON 50 2016, 159).

In short, from a legal and regulatory perspective, both COAs – self-defence and surrender – were available. Sovereign immunity and the inherent right of self-defence provide legal justifications to repel the IRGC boarders with proportionate (i.e. lethal) force. However, it would be an overreach to suggest that Article II or the SROE impose a legal *obligation* to fight.

Here, as discussed in Chapter 1, we find an essential distinction between legality and morality. When deciding between taking an action that is legal and one that is illegal, the law is unequivocal. The law can also distinguish which of two *unlawful* acts is more egregious. The law offers no basis, however, for determining which of two *legal* actions is the right thing to do. In this case, having established that both COAs were lawful, legal analysis had run its course. The law had nothing more to say about what the OIC *ought* to do. For this, he would have to rely on moral analysis.

NE203: Ethics and Moral Reasoning for Naval Leaders

The 1990–91 Defense Authorization Act directed America's four service academies[6] to introduce ethics to their core curricula. Although several high-profile ethics cases in the military and service academies may have contributed to Congress's decision to act, the primary rationale went beyond merely responding to bad press. The action acknowledged the evolving nature of power in global affairs. Poor moral choices on the modern, digitally connected battlefield could turn tactical successes into strategic failures. That America's officer corps firmly grasped this fundamental fact was a national security imperative.

The original design of the Naval Academy's ethics course – *NE203: Ethics and Moral Reasoning for Naval Leaders* – tracked closely with the ethics curricula of other colleges and universities that offered survey courses in moral philosophy.[7] The syllabus was structured around four major themes. First, the students considered the sources of their moral obligations, paying particular attention to culture, religion, and the distinctive moral obligation members of the US Armed Forces incur when they take their oaths to 'support and defend' the US Constitution. The second broad theme focused on character and virtue, especially the virtue ethics of Aristotle and the Stoics. The syllabus then turned to the major Western traditions of moral reasoning, introducing students to utilitarianism, Kantian ethics, natural law, and theories of rights and justice. The course concluded with an introduction to the just war tradition.

With vanishingly few exceptions, student feedback gushed with superlatives. NE203 consistently ranked among the highest-rated courses at the school. Yet amid the accolades and ample evidence of learning, there were two recurring critiques from students and faculty alike. First, the course seemed to move too quickly from one abstruse moral theory to the next. As a result, some students felt lost and became frustrated when competing theories seemed to offer conflicting solutions. The second general critique was that the course's emphasis on theory resulted in underemphasizing character and virtue. With only 2 weeks committed to Aristotle and Stoic philosophy, there was little room in the syllabus to examine virtue itself, character traits like courage, humility, and pride that encounter severe tests in the career of a naval officer.

The Farsi Island case reveals the canniness of these critiques. The OIC had taken NE203 his Youngster (sophomore) year. Yet during the most consequential moral decision of his life, it seems unlikely that the course, as structured, significantly influenced his reasoning. Without question, NE203 had provided him with ample tools for moral analysis, but without a process for deploying those tools, he lacked the time or inclination to consider any of them. Even if the OIC could have hit a 'pause button' and carefully examined each of the dozen or so moral theories he had learned in NE203, the course offered no basis for evaluating the relative importance of competing moral factors. How should he weigh his obligation to defend his boat from illegal seizure against his duty to respect the inherent worth of the US and Iranian crewmen who would surely perish in a point-blank exchange of automatic gunfire?

Prompted by insightful student feedback and similar observations from a 2017 visiting committee, department leadership decided to take action to shore up this otherwise exceptional course. The ethics faculty designed an experimental curriculum that ran parallel with the original design. The new curriculum responded to both concerns discussed earlier. First, it expanded the treatment of character and virtue from 2 to 5 weeks. The change permitted deeper analysis into specific character traits and created space for purposefully practising virtue (see Good 2023). Second, the curriculum systematized the course's theoretical material by introducing an analytical framework, 'the moral deliberation roadmap' (henceforth 'the Roadmap'). The Roadmap integrates elements from each of the moral theories presented in the original curriculum, but rather than introducing a proliferation of competing moral theories, it offers one unified, orderly process of moral deliberation.

After eight semesters of experimentation, USNA leadership decided to adopt the new ethics curriculum.

Moral Perception[8]

The remainder of this chapter will focus on the new curriculum's principal innovation: the Moral Deliberation Roadmap. First, however, we must address an important preliminary question: how do we know when it's time to consult the Roadmap? We don't move through the physical world with a map constantly before our faces. Nor should we navigate the moral world consulting Kant, Mill, and Aristotle on every decision.

Moral perception is the capacity to discern the morally relevant features of the world around us. A critical function of moral perception is alerting us to situations that warrant intentional moral reasoning. This section suggests three orienting questions that can sharpen moral perception.[9]

Question 1: Do I Recognize This Moral Terrain?

We make countless decisions every day, and many of these have significant moral content. They test our commitment to values we admire. They potentially harm or help others. Moral choices are all around us. When we smile and say good morning to co-workers even though we're feeling grumpy after a particularly awful commute, we have made a moral choice. We have made a moral choice when we walk past a colleague's empty and unlocked office and 'choose' not to empty the wallet sitting unguarded on his desk.

These are routine moral choices. By 'routine', I'm not suggesting they're unimportant. Indeed, they're supremely important. Everyday acts of kindness and civility – caring for our children and elderly parents, respecting the property of others, honouring widely embraced values like honesty and fair play – these are the choices that constitute the fabric of our moral universe. I characterize them as routine because we make them intuitively, without the need to engage in deliberate reasoning. We've learned to recognize patterns in the moral terrain and evolved habitual responses to them. Whenever we perceive a familiar pattern, we reflexively recognize our obligations and either act according to our moral habits or feel bad when we fail to do so.

Sometimes, however, the course of our routine decision-making is interrupted by a situation for which our settled moral habits are inadequate. An essential competency of moral perception is the aptitude for detecting these interruptions, for sensing that the moral terrain has become unfamiliar. If we correctly perceive a novel pattern, we can choose to slow down (if slowing down is an option) and, before acting, attempt to catalogue as best we can the relevant features of the new moral terrain. If, however, our moral perception fails us, we drive on, blithely relying on old maps to negotiate unmapped territory.

So, how do we know we've passed into unfamiliar moral terrain? There are cues. Simply the novelty of a situation is an important cue. 'Huh, I've never seen that before' is a good indication that our compendium of moral habits will not be reliable. Another cue is the recognition of conflicting moral obligations. Anyone who has ever told a lie to cover up the misdeed of a friend has experienced this. We can be honest, or we can be loyal. We can't be both. Finally, we have reason to suspect we're on unsteady moral ground when we just *feel* it; the hair on the back of our necks stands up, our stomachs churn, and our throats tighten. We argue in Chapter 1 that intuition is an unreliable moral arbiter. Intuition can, however, play a decisive role in alerting us to the need for moral arbitration; we ignore or undervalue intuition as an instrument of moral perception to our detriment.

In a 2006 talk at the US Naval Academy, Philosopher Rushworth Kidder described the boundary distinguishing routine choices from those that are morally ambiguous as a 'moral thermocline'.[10] He explained that water near the ocean floor is colder than that at the surface, but it doesn't grow colder gradually; it layers. A thermocline is a distinct boundary between the warm water above and the colder water below. A diver descending in a column of water may observe initially that the water temperature is relatively constant. Then, at a depth determined by temperature, density, water clarity, and other factors, she crosses the thermocline, feels the water temperature drop, and wishes she had worn a thicker wetsuit. The diver cannot *see* the thermocline – water is just as colourless below the thermocline as it is above – she *feels* it (Kidder 2006).

Kidder's moral thermocline works in the same way. Just as water is colourless above and below a physical thermocline, the moral landscape may appear to our rational mind as unchanged from Time-1 to Time-2. Yet we *feel* that something has changed. That weird feeling in the pit of the stomach signals a pre-reflective apprehension that our Time-1 moral habits may not be helpful at Time-2; they may even make the situation worse. During his discussion in Annapolis, Kidder advised the Midshipmen that they should learn to trust the moral thermocline. Whenever they feel that they've crossed it, they should pause, gather data, and give their analytic mind a moment to catch up with their gut. Excellent advice.[11]

Question 2: How Much Time Do I Have?

Time is an external factor that has an especially caustic effect on moral perception. In Chapter 1, we introduce an impressive body of experimental research that confirms this. Darley and Batson's 1973 experiment, for example, convincingly demonstrated that perceived urgency significantly influenced whether a Princeton seminarian would stop to assist a role-player slumped over on the ground or, pressed for time and oblivious to the evident suffering of the role-player, step over him. Similarly, Paxton *et al.* (2012) offer persuasive experimental evidence that as few as 2 minutes of forced reflection can radically change the moral choices we make.

Tragically, experimental findings like these have been reproduced outside the laboratory. On 9 February 2001, for example, a *Los Angeles* class submarine, *USS Greeneville*, was making a short transit outside of Pearl Harbor to demonstrate the sub's capabilities to a group of distinguished visitors (DVs). After morning events and lunch ran long, the Commanding Officer felt pressure to get the DV visit back on schedule. He started cutting corners. As *Greeneville* prepared to demonstrate an emergency ballast blow (a rapid surfacing manoeuvre that predictably earns DV oohs and ahhs), he decided to make up some time by waiving routine safety protocols (such as a periscope sweep for surface contacts). As a result of that moral choice, *Greeneville* surfaced underneath the hull of the *Ehime Maru*, a Japanese fishing vessel on a training voyage for high school students. *Ehime Maru* sank in minutes. Nine people, including four teens, were lost. 'A principal cause of the collision' concluded the Court of Inquiry convened to investigate the incident 'was an artificial urgency created by the CO' (Nathman *et al*. 2001, 103).

When we feel time pressing in on us, we alter our decision-making strategies. We limit our search for information, privilege data confirming our hasty conclusions, and discount facts contradicting them (Wildman 2011, 73–82). We forego deliberate moral reasoning and rush to judgement. Sometimes the time crunch is real. Typically, however, we have more time than we think. Before reaching for the Roadmap, therefore, it's prudent to ask, 'how much time do I have' for deliberation. If the answer is two seconds, so be it; making decisions with limited information is a necessary skill for military leaders. But if the answer is two minutes, two hours, or two days, we should take that time. 'The decisions we make on the cold side of the moral thermocline are too consequential to rush unnecessarily' (Herbert 2021, 58).

Question 3: Do I Already Know the Right Thing to Do?

Not all moral problems are moral dilemmas. Moral dilemmas arise when choosing between two or more COAs requires us to compromise one set of principles to uphold another. For example, a ship's commanding officer ought to be just, and she ought to be merciful. When disciplining a crewmember for an infraction, however, she often must choose which virtue to enact and which to compromise. Sending a good-order-and-discipline message to her crew may require her to favour justice over mercy. For moral dilemmas, the difficulty lies in deciding which moral principle we ought to validate and which to jilt.

Some moral problems aren't dilemmas at all; they're *tests of integrity*. When we confront a test of integrity, 'the difficulty lies not in deciding what the right thing to do is, but in actually doing it' (Coleman 2012, 5). Integrity tests are often confused with moral dilemmas, but they are a distinct class of ethical problems. Upholding one set of moral principles (e.g. fidelity to one's spouse) doesn't require sacrificing another set of moral principles, it requires eschewing something we may desire (e.g. pursuing an extramarital affair).

The value in asking Question 3 is that decision-making strategies diverge on this question. The Roadmap is a framework for working through moral dilemmas. It has little to say about resolving integrity tests. Moral deliberation for tests of integrity is simple because we already know the right thing to do. Given a choice between an act that is morally permissible and one that is not, we always ought to choose the former. Just do the right thing. Simple.

Of course, simple is rarely easy. Because we're human, knowing the right thing to do and doing the right thing are wildly distinct enterprises. Desire can be intense. Moral temptation is real. We've all failed integrity tests, and more failures lie ahead for most of us. Retired US Navy Admiral Bill McRaven, former Commander of

USSOCOM, offered the Midshipmen sagacious advice during a 2013 Forrestal Lecture at the US Naval Academy:

> Don't compromise your integrity . . . too often. The fact is, you *will* compromise your integrity. Why? Because you're human, and sometimes, the nature of self-preservation is too strong to overcome our sense of professional integrity. When that happens, it should make you sick to your stomach. You should question your moral compass, your parentage, your future as a naval officer, and your right to exist on this planet. And then, you should understand that you are not perfect and hope and pray that it never happens again. But it will. And each time you should feel sick. When you *stop* feeling sick for compromising your integrity, then it's time to leave the Navy.
>
> Try not to compromise your integrity.

Moral perception precedes moral deliberation. Without an account of moral perception, the Roadmap is an incomplete framework. This section has suggested three questions to focus moral perception and answer the broader question, 'is it time to pull out the Roadmap'?

In the Farsi Island case, it's likely that the OIC's answer to Question 1 would have been 'no'; he should quickly have recognized he was on unfamiliar moral terrain (a stark understatement). When the Iranian vessels closing his position removed the gun covers from their deck-mounted weapons, his intuition assuredly alerted him that he had crossed the moral thermocline. If his rational mind managed to catch up, he'd realize why his stomach was churning. He would have to choose between conflicting moral obligations: his obligation to respect human life and his duty to defend his craft from unlawful seizure.

In response to Question 2, it's arguable whether the OIC had sufficient time for intentional moral deliberation. Given his education and training in moral reasoning, it seems unlikely. The OIC was not introduced to the Roadmap as a Midshipman. A 'toolkit' containing a dozen or so ethical theories would have been of little use in the few short minutes available for deliberation. If, however, the OIC had access to a unified, orderly, and well-rehearsed process (i.e. the Roadmap) then I believe the OIC would have had sufficient time for moral deliberation. As the Paxton *et al.* (2012) experiment suggests, even a hasty (120 seconds) analysis can positively influence moral judgement.

In a wildly unscientific survey, I queried the Naval Academy's ethics faculty and found disagreement regarding Question 3, whether the OIC was confronting a no-kidding moral dilemma or merely a test of integrity. One faculty member maintained that the case did not involve a moral dilemma. The OIC, he insisted, was morally required to surrender. Several others similarly agreed that there was no dilemma but for the *opposite* reason: the OIC had a moral obligation to fight. Most of the ethics faculty, myself included, were satisfied that the Farsi Island incident qualified as an especially thorny moral dilemma and, therefore, a most appropriate case study for the Roadmap.

The Moral Deliberation Roadmap[12]

The fundamental intuition of the Moral Deliberation Roadmap is that regardless of the moral problem we face, there are four *moral factors* that are sometimes in conflict but always in play: constraints, consequences, special obligations, and character. When consulted in this order, these moral factors clarify and declutter the domain of actions under consideration. The Roadmap specifies which acts we can, can't, or must take and provides guidelines for adjudication when two or more factors seem to offer conflicting guidance.

Although the Roadmap asks us to examine moral dilemmas through four analytical lenses, it's helpful to consider the framework as a bifurcated process (see Figure 3.1). The Roadmap's upper-tier factors, constraints and consequences, illuminate our *general* moral obligations. These are duties that apply to all people

APPLIES TO **ALL** PEOPLE

1. Constraints

Do considerations of dignity, respect, rights, or justice require us to act in one way or another?

2. Consequences

What is going to bring about the best outcomes for everyone?

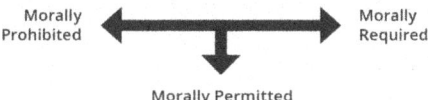

APPLIES TO **SPECIFIC** PEOPLE

3. Special Obligations

Does anything about the particular situation place you under an obligation others would not have in similar circumstances?

4. Character

How would this action affect your character? What's the action that an exemplary virtuous person would perform in this situation?

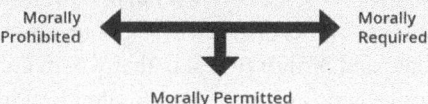

FIGURE 3.1   The Moral Deliberation Roadmap

Source: (Reprinted by permission of Roger Herbert and Marc LiVecche (eds) *Ethics and Moral Reasoning for Naval Leaders.* Pearson Education Inc.)

facing a similar situation, regardless of the decider's identity or background. After eliminating COAs that are 'morally prohibited' (or at least sending them to the parking lot from which few COAs return), we then submit the surviving COAs to the lower tier of the Roadmap. The lower-tier factors, special obligations and character, apply to specific people. Here we consider essential elements of the actor's biography: societal roles, personal relationships, promises made, and values embraced. Finally, after again eliminating any morally prohibited actions, we either take the action that is 'morally required'[13] or select from a now-reduced set of COAs, confident that all remaining options are 'morally permissible'.

Constraints

Constraints, as the Roadmap uses the term, 'limit a person's actions in deference to the special nature of human beings' (Skerker 2023a). Most Western cultures (and, by extension, most Western militaries) embrace the idea that all persons, *as persons*, possess inherent dignity and worth. Some ground this assertion in religious traditions, regarding basic human worth as 'a gift given to us by Another' (Eberle 2023). For others, our worth accrues from uniquely human capacities. Unlike non-human animals, humans are capable of exercising rationality and autonomy. We can comprehend the moral universe and impose rules of behaviour on ourselves, which we are free to obey or ignore. But regardless of its source, writes Eberle (2023), basic human worth endows all persons with 'a distinctive, morally privileged status' that confers a 'familiar package of natural human rights'. Human rights, in other words, are entitlements. They do not depend on desert or merit. They accrue equally to all persons 'no matter who they are, no matter their community . . . no matter what' (Eberle 2023). They accrue also to those who routinely (even joyfully) deny them to others.[14]

Skerker (2023a) defines rights as 'expressions of autonomy in specific areas of thought, speech, or action'. We are free to act within the scope of a given right. Additionally, rights justify our expectations that others will refrain from acts that would limit that freedom. In other words, rights impose duties. When rights-violators disregard their duties, rights-holders have cause to protest, expect sympathy from others, resist rights-violators with proportionate force, request forceful intervention from third parties, and demand restitution (Skerker 2023a).

A constraints-oriented analysis seeks to establish our moral obligations primarily by illuminating the rights of those we involve in our actions (thereby illuminating our moral obligations) and prohibiting acts that would contravene those rights.

But how can we know when actions violate another's rights? There's little consensus regarding the extent of human rights. The 1948 Universal Declaration of Human Rights, for example, asserts 30 distinct human rights. Jeremy Bentham, on the other hand, famously decried natural rights as 'nonsense on stilts'.

According to Prussian philosopher Immanuel Kant (1724–1804), our moral obligations are discoverable through reason. We obey our moral duty when our actions accord with 'the supreme principle of morality', the categorical imperative, an absolute and unconditional moral command that applies to everyone, always (Kant 2012 [1785], 4:392). Kant sets forth his categorical imperative in several formulations, which he insists are merely variations on the same question. The USNA curriculum focuses on the first two. The first formulation, the Universal Law Formulation, forbids actions that would result in a logical contradiction if we adopted the maxim behind the act as a universal law. The second, the End-in-Itself Formulation, prohibits actions that treat others not as ends in themselves but merely as means to our ends.

Skerker (2023a) restyles the categorical imperative as a three-part test to determine whether a contemplated act would constitute a rights violation. The 'universalization test' aligns with the first formulation. After restating our intended action as a *maxim* – a general policy or plan of action (O'Neill 2016) – we ask whether the maxim would result in a contradiction if it were universalized. Skerker offers a helpful way to gauge this: 'would my action be successful even if everyone in the world had the same 'maxim', or plan of action, I have'? (Skerker 2023a). Lying, Skerker points out, would fail the universalization test because successful liars depend on their targets' expectations of the truth. If we were to universalize lying, no one would expect the truth, and the 'institution' of lying would collapse.

If my maxim survives the universalization test, I then subject it to the 'mere means test' based on Kant's second formulation (Skerker 2023a). We all use people as means. Students presumably use their ethics professors as a means to learn moral theory (or complete a credit requirement). By contrast, when we treat other rational and autonomous beings as a *mere* means, we use them for our benefit alone, with no regard to the interests or advantage of the 'things' we're using. To treat someone as an end is to recognize they too have goals and values, to acknowledge their humanity. Again, Skerker offers a helpful assessment instrument: 'if the recipient of the action could in principle consent to the action being done to him', then he is not being treated merely as a means. When our ethics professor shows up to class apparently of her own volition and enthusiastically starts her lecture, we can infer consent. If, however, she teaches in shackles and at gunpoint, she becomes a mere means to our learning, not an end in herself.

Skerker offers a third test. Less directly tethered to the categorical imperative, the 'due respect test' accounts for Skerker's observation that some conspicuous rights violations survive the first two tests. The due respect test, writes Skerker, 'requires us to answer the following question: am I responding to others in a way that gives due weight to their worth as human beings'? If we accept that all humans possess great, equal, and inalienable worth (Eberle 2023), then some actions are so egregious they do not merit analysis.

While ordering a gunfight with crew-served weapons at close range may seem, *prima facie*, to be an egregious violation of rights – surely this must violate someone's rights – the OIC would not be able to rule out this COA based on constraints alone. A restatement of the action as a maxim, 'I will defend myself and my crew with proportionate force whenever a threat is imminent or ongoing, and escape is impossible', is essentially the definition of legitimate defensive rights. Exercising defensive rights creates no contradiction when universalized, and we can infer 'consent in principle' to a proportionate response to aggression on both sides of any military-on-military engagement. One could argue, perhaps, that the order to defend RCB 802 would have failed the due respect test. The foreseeable outcome of that order would have been mutual slaughter, hardly respectful of basic human worth.

Nevertheless, both COAs, to fight and to surrender, must be considered morally permissible at this point of our analysis. This verdict is based primarily on a crucial caveat to constraints-based reasoning: rights can be forfeited or waived. An actor *forfeits* a right by violating the rights of another. When the IRGC threatened imminent force, they forfeited the right not to be harmed. If the OIC had ordered a lethal response, he would not have violated the Iranians' rights. Nor, because of a similar caveat, would he have violated the rights of his own crew. One *waives* a right by consent. Just as boxer Joe Frazier waived his right not to be punched in the face when he entered the ring with Mohammed Ali, the crewmembers of RCB 802 waived – through their oaths of enlistment – their rights not to be placed in harm's way.

Using the language of just war theory, engaging the Iranian gunboats with lethal force would constitute a *discriminate* use of force. The *jus in bello* criterion of discrimination requires soldiers in combat to discriminate between combatants and noncombatants; only the former may be intentionally harmed. Even though they are uniformed and armed, members of the IRGC patrolling Iranian waters in peacetime are non-combatants. But the moment they threatened the Americans with imminent lethal force, they forfeited their non-combatant status and became legitimate targets.[15]

Before turning to our second moral factor, consequences, a final comment regarding constraints is appropriate. The USNA ethics curriculum considers two additional concepts under the rubric of constraints that I have not yet mentioned. First, there is another caveat that is not relevant to the Farsi Island case but critically important when considering constraints: the principle *of double effect*. Although intentional rights violations are almost never permissible, one may *foreseeably* violate the rights of another when 'there will be a simultaneous and inseparable good and bad effect as a result of one's normally permissible action' (Skerker 2023b). In other words, the bad effect is unavoidable, it is not the means to achieve the good effect, and the harm of the bad effect is offset by the good produced by the good effect. The good effect of eliminating an enemy headquarters may justify foreseeably, though unintentionally, killing the janitorial staff.

Second, considerations of rights do not exhaust the category of constraints. Respect for human dignity and worth also demands that we treat people fairly. Although not relevant to this case, constraints-based reasoning must also include considerations of distributive justice (the fair distribution of the benefits and burdens of a community) and retributive justice (fair punishment for past misdeeds).

Consequences

Philosopher David Rodin's (2002, 25) description of rights offers a helpful analogy for understanding the relationship between constraints and the rest of the Roadmap. Rather than likening a right to a trump card, a moral consideration that beats all others, Rodin prefers a 'breakwater' analogy, a structure that protects in all but extraordinary circumstances. The function of a right, he suggests, is to,

> erect a normative barrier against the infringement of individual interests and liberties. It is in the nature of this barrier to be sufficient to defeat the great majority of competing claims. However, it is always conceivable that circumstances will arise in which an individual's right is simply overwhelmed by the gravity of competing moral considerations.

The Roadmap starts with constraints because human rights, dignity, and justice should generally override other moral factors, as Rodin's breakwater analogy suggests. If analysis of constraints yields the result that an act is morally prohibited, it can't be morally justified by good consequences unless those consequences are exceptionally good or bad.[16]

Yet consequences are vitally important, especially in the military profession. In my career as a naval officer, my seniors never graded me on my good intentions; they judged me on outcomes. So, after eliminating morally prohibited actions, it's most appropriate to turn to the foreseeable consequences of our actions.

Consequentialist reasoning comes naturally to us. Like other animals, humans are naturally disposed to pursue pleasure and avoid pain. Consequentialists maintain that this predilection is morally relevant or, for some, morally decisive. There are several forms of consequences-based theories that riff on this intuition. What they share is that an action is right or wrong based on foreseeable outcomes; the principles behind the act are irrelevant. Utilitarianism, the consequentialist theory most closely aligned with USNA's curriculum, holds that an act is right or wrong based on the 'principle of utility'. We ought to choose the action that generates the greatest *net happiness*: the overall pleasure pumped into the world by a given action minus the overall pain. Importantly, utilitarianism is a non-egoistic theory. 'My/our' happiness counts but no more or less than 'yours/theirs'.

While utilitarianism is a simple theory, determining net happiness is a demanding enterprise. It involves consideration of at least four questions: 1) what are the short-term consequences? 2) what are the long-term consequences (including the precedent a given action sets when)? 3) who is affected and how? 4) how likely is each outcome (Skerker 2023c)? A common critique of utilitarianism is that answering these questions requires omniscience. This is unjustified. Although utilitarianism sets a high epistemic bar, it asks only that we do our best given the information and time available.

In the Farsi Island case, I have argued that the first moral factor, constraints, neither morally prohibits nor requires either COA (fight or surrender). Both COAs, therefore, are candidates for consequentialist analysis. In the short term, mounting a defence would likely be bloody for both sides. It's conceivable that no one would walk away unharmed, leaving scores of people to grieve the dead and administer to the maimed. In the long term, the decision to fight would affirm the precedent that anyone attempting to violate US sovereignty or unjustly threaten Americans will pay a dear price. However, a bloody exchange could set off an escalatory spiral involving entire nations. The short-term consequences of surrendering are less clear. Surrender could result in a brief but embarrassing (for the crew, the Navy, the United States) visit to a temporary holding facility while the diplomats negotiate. Extended incarceration, torture, or even death are plausible outcomes, though less likely. The most likely long-term price is reputational; the precedent of American weakness may endanger other US service members.

In combat or situations like the Farsi Island encounter, analysis of consequences serves the purpose of the proportionality and necessity criteria of the *jus in bello* convention in the just war tradition. If the foreseeable outcomes of an action cause more harm than good, then it is a *disproportionate* use of force. If a military objective can be achieved at a given level of violence, then choosing to apply a greater degree of violence is an *unnecessary* use of force.

At this point in the analysis, smart people may still disagree on what the OIC ought to do. However, after considering constraints and consequences – the two moral factors that apply to all people facing similar circumstances – the moral picture seems to be coming into focus. The OIC can now consider the moral factors specific to him and this crisis.

Special Obligations

Special obligations are duties we incur by virtue of some aspect of our identities or life stories. Roles we've assumed, relationships we've established, promises we've made, and other idiosyncratic circumstances may impose obligations that others in similar situations do not bear. Special obligations serve two purposes in moral deliberation. First, they can act as 'tie breakers'.[17] If two actions are morally permissible, allowing special obligations to decide the winner is appropriate. Special obligations can also play a more substantial role by recasting an action that is merely permissible into one that is morally prohibited or required. They can also render morally blameworthy acts that would otherwise be considered morally praiseworthy.

Philosopher Michael Sandel (2009, 208–243) distinguishes two categories of special obligation. He calls the first 'voluntary obligations'. We incur voluntary obligations by consent. If I sign a contract, make a promise, or take an oath, I consent to be constrained by that contract, promise, or oath. For example, it's morally permissible, even laudable, for a noncombatant to flee a gunfight. But because of her oath, a soldier is morally (and legally) obligated to run toward the sound of the guns if ordered to do so. Sandel's second category includes 'obligations of solidarity'. Because families, communities, platoons, and nations share a common history and typically a common future, we incur a duty to pay special attention to the needs of family members (regardless of past wrongs), neighbours (even if we find them obnoxious), colleagues (the overachievers and the slackers), and compatriots (though they may be strangers). If a father rescues a stranger's child from drowning, his action is morally praiseworthy. If, however, his own child is also drowning, and he can only rescue one, we would find it morally questionable to say the least if he chose to save the stranger's child and allow his own to drown.

In war, special obligations are in constant tension with general obligations. Captain Rick Rubel, the founding Course Director of NE203, branded this tension the '3-Way Moral Problem of War'. Every military leader has a special obligation (voluntary obligation) to accomplish the assigned mission and a special obligation (obligation of solidarity) to the welfare of the troops. These two commitments are themselves often in tension, even in peacetime. During times of war, however, they are both in tension with a general obligation: the safety of noncombatants, those who have done nothing to forfeit or waive their rights not to be harmed. When special obligations conflict with general obligations, Rodin's breakwater analogy described earlier pertains. Except in extraordinary situations, general obligations trump special obligations. Chris Eberle expresses this tension well in an email he shared with the Naval Academy's ethics faculty in 2020.

> It's the combination of dignity, rights, justice, and consequences that trump special obligations, ordinarily, not always. So, when all those considerations align, we should be very, very wary of adhering to our special obligations, though we will be powerfully tempted to do so.

How might considerations of special obligations have affected the OIC's decisions off the coast of Farsi Island? The OIC's voluntary obligations, which he incurred when he took his oath, should bias his decisions toward completing the mission: transiting from Kuwait Naval Base to NSA Bahrain. It's unclear, however, which COA would have best served that obligation. Had the OIC ordered his crew to fire, it's unlikely that RCB 802 would make it to Bahrain, and certainly not in a condition for follow-on tasking. The Code of Conduct's never-surrender ethos also confers a voluntary obligation on service members. Still, that ethos, as discussed, is an awkward fit for the pseudo-war that defined America's relationship with Iran in 2016. The COA advised by the OIC's obligations of solidarity seems more readily discernible. Although surrendering his crew would put them at risk of harm, ordering a gunfight at that range from a boat unable to manoeuvre would have assured grave harm to the bodies, minds, and souls of individuals to whom the OIC owed a special obligation.

Character

During the summer of 2019, USNA's LEL Department launched 'Project Babel', a contingent of faculty voiced concern that our students were badly overusing or misapplying terms that are essential to the profession of arms. Project Babel set out to refine definitions; character was among the terms we considered. After a summer of research, Team Babel settled on a simple definition: character is 'the sum of our moral habits'.

Thinking of character as a big sack of habits helps us understand why the Roadmap includes character among its four moral factors. A character-as-habit definition prompts us to consider how habits, in general, are formed, reinforced, and broken. Although habits, by definition, are non-deliberative actions, they are formed quite deliberately. We make intentional choices that become tendencies if we consistently make the same choice in comparable situations. They become habits when we don't have to choose at all. We just act. We weaken our habits by acting contrary to our habituated responses and break them by consistently making the contrary choice, thus forming a new habit.

Character development proceeds in much the same way. Like habits in general, we affect our character – our sack of moral habits – every time we choose. Each moral choice we make either lays the foundation for a new moral habit, reinforces existing habits (the bad ones and the good), or, when we act contrary to our moral habits, weakens them (for better or worse). Most moral choices we make affect our character imperceptibly. Sometimes, however, changes are profound.

What this suggests is that if an action reinforces a character trait we value, we have good reason to take that action. We also have good reason to avoid acting contrary to our admirable moral habits. For example, if we admire courage, then we should consistently make choices that the virtue of courage demands. We should ask ourselves what action a courageous person would take in the situation at hand and act accordingly. Because we degrade our character when we make a choice that is cowardly (a deficiency of courage) or reckless (an excess of courage), we should always stand up to bullies and say 'no' when dared to pick up venomous snakes.

The IRGC's assault on RCB 802 presented the OIC with a formidable and unwelcome character-building (or character-degrading) opportunity. Pride, humility, and temperance – virtues to which USNA's current syllabus commits significant attention – were all tested that afternoon. It was the OIC's courage, however, that faced the severest test. Turning again to Project Babel, courage is 'doing the right, just, or noble thing despite known risk, and typically despite fear'.[18] By this definition, ordering his crew to repel the Iranian assault would have demanded extraordinary physical courage. Although it's arguable whether this COA was *the* right and just action, we have established that defending the boat was at least *a* right and just action. And certainly, there's nobility in using proportionate force to protect one's rights against an egregious rights violation. Finally, the OIC could have had no doubt that the risk of bodily harm to his crew and himself was grievous.

The alternative COA, surrendering his crew, would also require physical courage; captivity involves the risk of deprivation or torture. But surrendering would have presented an even greater test of *moral* courage. We display moral courage when we act despite social, political, economic, or psychological risks. When the

OIC surrendered his crew, he assumed risk in each of these categories. Would he forfeit his career? Would he be judged a coward?

One final thought before leaving the topic of character. The risk of character degradation is especially pronounced in war. Jonathan Shay's influential book, *Achilles in Vietnam* (1994), introduced the idea that PTSD is not the only 'unseen wound' in combat; some injuries are *moral*. Shay (2012, 57) describes moral injury as a 'soul wound'. Brett Litz and his colleagues (2009, 700) define moral injury as a loss of trust, profound guilt, or degradation of virtue resulting from 'perpetrating, bearing witness to, or learning about acts that transgress deeply held moral beliefs and expectations'. Moral injury varies in severity. Ned Dobos (2015, 126) distinguishes two categories of moral injury: 'moral trauma' and 'moral degradation'. Dobos explains, 'A soldier who suffers moral trauma *feels like* a bad person. A soldier [who] suffers moral degradation *is* a bad person'. Physical injury can leave a victim permanently deformed. Similarly, moral injury is a loss of functionality that 'can permanently warp a soldier's capacity to experience empathy for the suffering of others or distinguish right from wrong or feel remorse for genuinely wrongful acts' (Baker *et al.* 2023).

Conclusion

When confronted with moral dilemmas, particularly high-stakes moral dilemmas that are routine in war, we feel profoundly alone. Yet we're never alone. As it turns out, humans have been thinking about how to make good moral choices, how to become people of good character, and how to fight and win with honour – and return from war as whole human beings – for millennia.

Accessing the wisdom of the ages, however, is often most challenging when it is most critically needed, at moments of high stress and consequence. This is the purpose of the Moral Deliberation Roadmap. The US Naval Academy's ethics faculty designed a unified, orderly process for moral deliberation, one that their graduates can employ, even at the speed of war.

When I taught this course, my students would reliably ask me after we had worked through a particularly vexing case study, 'What would you have done, Captain'? My responses tended to be rather slippery. I wanted my students to think about these cases long after the course, and 'solutions' offered by senior officers can shut down curiosity. So, I would stroke my beard and channel my inner Socrates, throwing questions back to them until either they lost interest in the exercise, or the bell rang.

I'll break with this annoying convention here and offer my opinion on the OIC's decision off Farsi Island. While the OIC benefited from a superb ethics course when he was a Midshipman, he was never introduced to the Moral Deliberation Roadmap, which had not yet made its NE203 debut when he took the course. Nevertheless, the OIC, in my opinion, got it exactly right. Although both options – surrendering and fighting – were morally permissible, the cost of a point-blank firefight with heavy weapons would have been exorbitant. It would have been a price worth paying in wartime, but not on that afternoon. Furthermore, a bloody encounter in Iranian territorial seas is the type of incident that could push a pseudo-war into something much more costly for both states. It's unclear which COA would best fulfil the OIC's special obligation to accomplish his mission, but his special obligation to bring his troops home, with body and soul intact, demanded his surrender to the Iranians.

The OIC's decision under extreme stress is a testament to his moral intuition and character. Mostly, however, his decision that afternoon was a testament to his courage. It's the decision I hope I would have made.

Notes

1 The Riverine Command Boat is a 53-foot vessel used primarily for port security. It has a crew of four to six, a max speed of 40+ knots, and a range of 250+ nautical miles. It has four weapons mounts that accommodate .50 calibre, 7.62mm, or 40mm machine guns.

2 My summary of the Farsi Island encounter relies primarily on Commander, Destroyer Squadron Fifty's (COMDESRON 50) command investigation and an email dialogue I enjoyed with the OIC, whose candour and continued concern for the wellbeing of the sailors with whom he served is inspiring. My chapter does not consider events (or their legal/moral implications) during the period in which the crew was in Iranian custody.

3 The rationale given for this decision was that the first-class petty officer had more experience than the OIC. This is an unconventional decision. It is not uncommon for NCOs to have more experience than the junior officers who lead them.

4 LEL Department offers four required courses: a leadership course during both Plebe (freshman) and Second Class (junior) years, ethics as Youngsters (sophomores), and an introduction to relevant law during their Firstie (senior) year.

5 Non-innocent activities include threatening or using force; firing weapons; collecting intelligence; promulgating propaganda; launching or recovering aircraft; transferring cargo, currency, or persons; polluting; fishing; conducting unauthorized research; and interfering with communications.

6 The United States Military Academy (West Point, New York), United States Naval Academy (Annapolis, Maryland), United States Air Force Academy (Colorado Springs, Colorado), United States Coast Guard Academy (New London, Connecticut).

7 Although USNA's syllabus was unremarkable, the course's overall framework was. As an elegant solution to understaffing, the lead course designers – retired Navy Captain Rick Rubel, USNA's Distinguished Military Professor of Ethics, and Professor George Lucas, one of the world's foremost voices in Military Ethics – paired PhD philosophers with senior naval officers. The result of the Rubel-Lucas, team-teaching model was a course in theoretical and applied military ethics that was unrivalled.

8 This section is an expanded version of my seminar notes, incorporating insights from multiple sources, including lectures by Ed Barrett, Chris Eberle, Marcus Hedahl, Pierce Randall, and Michael Skerker.

9 I introduce these questions in Herbert 2021, 56–60.

10 To my knowledge, Kidder's excellent metaphor does not appear in any of his published works.

11 Commenting in an email to me on Kidder's metaphor, my successor at the US Naval Academy, Mike Good, emphasized the role that emotions play in moral perception. 'The sensitivity to the temperature change will depend on one's habituated perceptual capacities . . . particularly in her emotions'. The moral thermocline alerts us to 'to something

different or something to which you need to attend. Emotions are those alerts in moral life'.

12 This section is an expanded version of my seminar notes, incorporating insights from multiple sources, including Eberle 2023; Skerker 2023a, 2023b; and lectures presented by Ed Barrett, Chris Eberle, Marcus Hedahl, Pierce Randall, and Michael Skerker.

13 An action that is 'morally required' is a 'positive' or 'imperfect' duty. Analysis using the Roadmap is more likely to yield a verdict of 'morally prohibited' (a 'negative' or 'perfect' duty) or merely 'morally permissible'. A perfect duty, for Kant, is universally applicable to all people. His imperfect duties are those that we have to some people some of the time, much like the Roadmap's special obligations.

14 Thanks to Bob Schoultz for this observation.

15 Thanks to Marcus Hedahl for his efforts to emphasize the congruency between the Roadmap and traditional terminology dealing with the ethics of war.

16 The default position that principles typically trump consequences is, of course, arguable. The academics and naval officers who crafted the Moral Deliberation Roadmap agreed that the inherent bias toward consequentialist reasoning in the military service can lead to immoral outcomes. Sequencing deontological considerations prior to consequentialist factors mitigates the risks of the consequentialist bias.

17 Thanks to Michael Skerker for this insight.

18 Four members of the USNA faculty focused on defining courage for Project Babel: Major Andy Hiller (US Marine Corps), Lieutenant Commander Danielle Litchford (US Navy, retired), Colonel Maria 'MJ' Pallotta (US Marine Corps Reserve), and me.

References

Baker, Deane-Peter, Roger Herbert, and David Whetham. 2023. *The Ethics of Special Ops: Raids, Reconnaissance, Recoveries, and Rebels*. Cambridge University Press.

Coleman, Stephen. 2012. *Military Ethics: An Introduction with Case Studies*. Oxford University Press.

Commander, Destroyer Squadron Fifty (COMDESRON 50). 2016. Command Investigation to Inquire into Incident in the Vicinity of Farsi Island Involving Two Riverine Command Boats (RCB 802 and RCB 905) on or About 12 January 2016.

Darley, John M., and C. Daniel Batson. 1973. '"From Jerusalem to Jericho": A Study of Situational and Dispositional Variables in Helping Behavior.' *Journal of Personality and Social Psychology* 27(1), 100.

Dobos, Ned. 2015. 'Moral Trauma and Moral Degradation' in Tome Frame (ed.) *Moral Injury: Unseen Wounds in an Age of Barbarism*. UNSW Press.

Eberle, Christopher J. 2023. 'Human Tribalism, War, and Human Dignity' in Roger Herbert and Marc LiVecche (eds) *Ethics and Moral Reasoning for Naval Leaders*. Pearson Education.

Good, Michael. 2023. 'Disciplines for the Moral Life' in Roger Herbert and Marc LiVecche (eds) *Ethics and Moral Reasoning for Naval Leaders*. Pearson Education.

Herbert, Roger. 2021. 'Moral Reasoning in Seven Questions.' *U.S. Naval Institute Proceedings* 147(5), 56–60.

Kant, Immanuel. 2012. *Kant: Groundwork of the Metaphysics of Morals*. Cambridge University Press.

Kidder, Rushworth. 2006. *Remarks during the US Naval Academy Leadership Conference*. United States Naval Academy.

Litz, Brett T., Nathan Stein, Eileen Delaney, Leslie Lebowitz, William P. Nash, Caroline Silva, and Shira Maguen. 2009. 'Moral Injury and Moral Repair in War Veterans:

A Preliminary Model and Intervention Strategy.' *Clinical Psychology Review* 29(8), 695–706.

McRaven, William H. 2013. *Forrestal Lecture*. United States Naval Academy.

Nathman, John B., Isamu Ozawa, David M. Stone, and Paul F. Sullivan. 2001. *Court of Inquiry into the Circumstances Surrounding the Collision between USS Greeneville (SSN 772) and Japanese M/V Ehime Maru That Occurred Off the Coast of Oahu, Hawaii on 9 February 2001*. US Navy.

O'Neill, Onora. 2016. 'A Simplified Account of Kant's Ethics' in Larry May and Jill B. Delston (eds) *Applied Ethics: A Multicultural Approach*. Routledge, 16–20.

Paxton, Joseph M., Leo Ungar, and Joshua D. Greene. 2012. 'Reflection and Reasoning in Moral Judgment.' *Cognitive Science* 36(1), 163–177.

Rodin, David. 2002. *War and Self-Defense*. Oxford University Press.

Sandel, Michael J. 2009. *Justice: What's the Right Thing to Do?* Farrar, Straus, and Giroux.

Shay, Jonathan. 1994. *Achilles in Vietnam: Combat Trauma and the Undoing of Character*. Scribner.

Shay, Jonathan. 2012. 'Moral Injury.' *Intertexts* 16(1), 57–66.

Skerker, Michael. 2023a. 'Constraints' in Roger Herbert and Marc LiVecche (eds) *Ethics and Moral Reasoning for Naval Leaders*. Pearson Education.

Skerker, Michael. 2023b. 'Constraints Complications: Waiving, Forfeiture, Double Effect, and Justice' in Roger Herbert and Marc LiVecche (eds) *Ethics and Moral Reasoning for Naval Leaders*. Pearson Education.

Skerker, Michael. 2023c. 'Minimizing Harms: An Introduction to Consequentialist Reasoning' in Roger Herbert and Marc LiVecche (eds) *Ethics and Moral Reasoning for Naval Leaders*. Pearson Education.

Wildman, Jessica L. 2011. 'Trust in Swift Starting Action Teams: Critical Considerations' in Neville Stanton (ed.) *Trust in Military Teams*. Ashgate Publishing, 73–82.

4

A QUASI-UTILITARIAN APPROACH TO DECISION-MAKING IN WAR

Iain King

An elderly Iraqi cleaner jogs towards the concrete shelter, a group of European soldiers in their gym kits run to their body armour, and the driver of an armoured vehicle locks the doors and reverses to safety. . . . Fortunately, the alarm blaring out over the base in Baghdad soon stops. It was a false alarm: there is no incoming ordnance, and nobody will be hurt. Not today.

Incidents like this are becoming much rarer in Iraq, but there are still people nearby with a will to inflict harm.

Causing pain is fundamental to war. Violent conflict is about doing things that another person fundamentally does not want done.

How can we justify that which is unwanted? Some historic justifications for war seem preposterous.[1] Given that war brings devastating impacts upon people and societies, the justifications for it must be thoroughly robust. And given that each war gives rise to so many grisly events and nasty choices, it is vital that there is a means to decide what armed servicemen and women, and their leaders, should do and should not do.

Sadly, there is no single, agreed means to determine which wars, and which acts of violence, are morally legitimate. That is largely because there is no single, agreed means to determine how right differs from wrong, and good from bad.

This chapter will show that it *is* possible to give clear ethical direction – moral guidance which is credible, coherent, and comprehensive. It will then apply this approach to give practical answers for humanity's most horrific dilemmas: those which arise around warfare.

DOI: 10.4324/9781003312925-5

Hard Wars, Hard Choices

Consider these six examples of real-life conundrums in warfare:

1 Major Howard is ordered to take a bridge from the enemy. The bridge is of crucial importance for his country's war effort, and he is lucky enough to have reliable and accurate intelligence about the enemy's presence in the area. How should he proceed?
2 Police Commander Trenton is responsible for preventing terrorism in her country. The terrorists believe they are fighting a war with legitimate political goals; Trenton regards them merely as criminals. What should her approach be?
3 Brigadier James is Commandant of a military academy, where young men and women start their military careers. What is the right way to prepare future senior officers?
4 Sven, a medic captured in the Bosnian war, is given a revolver and instructed to kill one of his fellow prisoners of war. If he does not, he is told ten POWs will be killed. Should he pull the trigger?
5 Private Wills initially enjoyed combat, but after two-and-a-half months of killing enemies with his bayonet, and seeing close friends die, he feels revulsion. He is sent into battle again. What should he do?
6 Captain Ivan is instructed to fire his artillery 20 miles west, to a location he thinks may be a school, and he has limited trust in his seniors. Should he fire as instructed?

Resolving these dilemmas requires not just an understanding of war. It requires an understanding of right and wrong. And that leads, naturally, to ethics.

This chapter will derive answers to these questions by establishing a system of ethics which is credible, coherent, and comprehensive. It will do this by exploring what statements of right and wrong mean; developing an evolutionary basis for our ethical sentiments; and then seeing how they can be rationalized to provide suitable advice for difficult situations. It will show that this advice is based around a so-called 'Help Principle', and a form of quasi-utilitarianism.[2] It will then apply it to the aforementioned situations, to show what needs to be done in each case. Finally, it will offer some general conclusions about decision-making in the traumatic confusion of war.

Generating Ethical Advice for War

War fighters require a theory of ethics which is credible enough to be trusted; comprehensive enough to cover the full range of dilemmas; and coherent, so that it gives clear rather than contradictory advice. To establish such a system of advice requires an understanding of what ethical propositions mean.

There is widespread disagreement about how, exactly, concepts like right and wrong operate. Questions about moral guidance rest on questions of meta-ethics – that is, 'the attempt to understand the metaphysical, epistemological, semantic, and psychological, presuppositions and commitments of moral thought, talk, and practice' (Sayre-McCord 2012). Metaethics 'takes a birds-eye view on the practice of ethics . . . (peering down) intently, trying to make sense of what is going on' (Fisher 2014, 2).

Sadly, like ethics, meta-ethics is passionately contested. But it is possible to establish an account of meta-ethics which is good enough to provide a foundation for ethics – and so, in turn, a credible basis for practical advice.

The most plausible accounts of how ethical instincts arose suggest ideas of right and wrong evolved in people because it confers an evolutionary advantage (Wielenberg 2016, 502–515) – that 'human morality is a distinct adaptation wrought by biological natural selection' (Joyce 2008, 214). The hypothesis is that communities of people who shared certain traits – such as being mutually supportive or brave in the face of outside threats – would survive and propagate; and eventually their members would outnumber and displace those in less 'moral' communities (Ibid.). Hence, the communities around today will tend to be those which developed social codes and norms which helped them survive in their environment over many generations.

Given that our environments are arbitrary,[3] and also that they differ around the world, this evolutionary account of how ethics emerged suggests our ethical attitudes have an arbitrary quality, too. But it is possible to accept the very credible hypothesis that our ethical attitudes are mostly the product of an amoral evolutionary process, while also accepting that we are unable to escape them. Knowing, say, my aversion to bayonetting babies, or to the slaughter of POWs, is the upshot of my ancestors' adaptation to arbitrary circumstances over millennia will not persuade me that bayonetting babies or war crimes are acceptable. We are entitled to accept some moral statements as facts, even though they may be inexplicable to us – what philosophers Searle and Anscombe called 'brute facts'.

Hence, we are forced to accept an apparent paradox in our ethical sentiments: ethical motivations feel real to us, and they compel us to behave in certain ways, despite strong evidence that the basis for them is arbitrary. We have good reasons to recognize that when ethical pushes and pulls guide us one way, it would have taken only an accident of culture or of our ancestor's evolutionary environment for them to be guiding us another. And yet, we cannot readily step outside the ethical framework these accidents have bestowed upon us (King 2014).

So, we can and do talk as if there are ethical facts – such as 'war crimes are wrong'; we really do accept these as facts, and regard them as legitimate as other facts (Ibid.), while simultaneously accepting these are of a different nature to all other facts we know. There is something odd about ethical facts: they are not like other 'real' things.[4] They are 'quasi-real'.

Our Instincts to be Ethical

Our ethical instincts often motivate us. Sometimes they are internalized, such as when we barely think before helping a wounded comrade; and at other times, they are the starting point for active deliberation, such as how our instinct to avoid harm to civilians affects military planning.

Our moral instincts can change over time, such as the recent shift in attitudes towards women serving in frontline military roles. But there is a natural speed limit to these changes (Blackburn 1998, 314).

Also, it is clear that contradictory ethical instincts exist within people as well as between them (Sartre 2007 [1946]). There is a tension between the urge to harm an enemy, and the will to be merciful. Determining how to resolve tensions between rival instincts is a key aspect of defining how to behave ethically (Ibid.). We expect ethical advice to be the same for all situations with identical moral elements (to proffer different advice for morally equivalent cases betrays a misunderstanding of what 'right' and 'wrong' mean). Ethical advice needs to be coherent; if ethical advice is incompatible with itself then it is no advice at all.

The fact that people are often conflicted about what they should do in testing situations means that, to establish a credible basis for ethical behaviour, we need a means for deciding between competing ethical drives.[5]

Coherence Versus Comprehensiveness

Several pre-existing ethical systems offer ways to remove the incoherence in ethical advice. Utilitarianism, for example, provides a single recommendation for every situation: choose whichever course will generate the greatest total happiness. Similarly, a Kantian perspective, which advises people to 'Act only according to that maxim by which you can at the same time will that it should become a universal law' (Korsgaard 1985, 24), should also offer coherent recommendations to any person with a clear view of what they believe the universal law should be. Virtue-based ethical systems which define a single characteristic as good[6] can promote a single and coherent course of action, too.

But in all these examples, coherence comes with a significant cost: each of these systems provides coherent responses by locating ethics in just one part of the decision-making process. A virtue theorist looks just to the character of the agent; a Kantian studies only their actions; while a utilitarian looks solely to the outcome they create. For all of them, coherence comes by tuning out any moral considerations which may jar with their single metric of good and bad.[7]

This means all mainstream normative theories provide unappealing advice for some scenarios. Utilitarians are accused of corrupting justice, because they would issue punishments without direct regard to any crimes committed, including penalties to innocent people if it was an effective deterrent (Haist 2008, 789). Kant's commitment to truth-telling is attacked because he would be compelled to inform a mad axe murderer where his intended victim was hiding: by ignoring consequences, Kant is complicit in an immoral outcome.[8] Similarly, a pure virtue theorist has to accept that, by locating goodness solely in character, their concept of a 'good person' allows someone to commit an endless string of bad deeds with nasty outcomes. 'There are certain features of morality which a virtue-based ethic either handles poorly or ignores entirely' (Louden 1984, 229).

In part to avoid these critiques, most modern approaches to ethics will take into account multiple factors. Individuals are expected to show good character (virtue ethics), to follow certain rules of behaviour (deontology) and to achieve positive results (consequentialism). But as soon as the ethical checklist has more than one item on it, there is a possibility – often a likelihood – that the different ethical considerations will provide conflicting advice, leading, once more, to incoherence.

So, the challenge is to define ethical behaviour in a way which takes in these divergent ethical perspectives while remaining coherent – that is, in a way which is able to arbitrate between different ethical considerations. That, in turn, requires a credible basis for ethics.

Constructing a Credible Basis for Ethical Decisions

Moral theorists have approached this problem before, and have usually sought to establish a common point of reference from which their ethical system can derive. John Rawls (1971), for example, constructed an imaginary 'Original Position' in which people were asked to decide principles on the allocation of rights and resources, and how a society should operate. John Stuart Mill tried to set out a foundational principle for his account of utilitarianism from the fact that individuals generally try to do what is best for themselves (Rosen 2005); his argument was that, if their decisions were made suitably impersonal, then people should choose what is best for everybody (Ibid.).

While noting that both Rawls' and Mill's approaches have been criticized (see, e.g., McCormick 1998, 748–750), they offer valuable insights. Both seek to bring out the apparent 'logical structure' of ethics (Grayling 2019, 449). And both take a minimalist approach: they distil ethics by stripping away elements which complicate or distort judgements. By adjusting both approaches to correct their mistakes, it is possible to derive a suitable foundational principle of ethics.

Several commentators suggested Rawls 'rigg(ed) the rules' (Rowland and Rowland 1995, 345) to generate his preferred outcome. When Rawls hypothesized people deciding principles behind a veil of ignorance, where he removed their knowledge about their religion, military affiliation, wealth, and education (so as to remove their biases on those issues) he allowed them to know they were risk-averse (Rawls 2009 [1971], 133–136). Applying the minimalism criteria mentioned earlier to strip away all that could be irrelevant to moral decision-making, it is appropriate to exclude risk aversion, too.

Rawls is guilty of biasing his people behind the veil of ignorance to favour the future over the past. The hypothetical contract derived in Rawls' original position shares out benefits in the future without reference to any previous promises or commitments. Rawls shares his 'future bias' with utilitarianism, and it does not survive the minimalism test (Ibid.). It must go, too.

Also, we need to clarify how Rawls' scenario should deal with secondary preferences: is a soldier admired by others in her unit, or on whom that unit depends, more entitled to benefits than a loner? Rawls obfuscated. Fortunately, the minimalism criteria provide an easy answer to this: benefits to someone should not count for more just because another person wants them to prosper. However, in part to deter those who wish harm to another, it is possible to derive a principle of reciprocity, which means benefits should be extended to people only in proportion to how far those people extend the principles to others (King 2008, 109–114).

Once these changes are made, in place of Rawls' 'Difference Principle' and other precepts, the essence of what they would adopt is summarized by the Help Principle:

'Help someone if your help is worth more to them than it is to you'.

This outcome is confirmed when adjustments are made to John Stuart Mill's argument for utilitarianism. John Stuart Mill sought to prove his theory, that 'actions are right in proportion as they tend to promote happiness, wrong as they tend to produce the reverse of happiness' (Mill 1991 [1863]), by arguing that individuals naturally choose what is best for themselves, and doing good requires a common rather than a selfish approach, so everyone should do what is best for everyone overall (Ibid.). Mill wrote, 'each person's happiness is a good to that person, and the general happiness, therefore, a good to the aggregate of all persons' (Ibid.).

Mill's proof fails (see Seth 1908). It has the same structure as 'each person chooses their own target in a battle', to 'therefore everyone should choose everyone's target in a battle': it conjures an 'everyone' which selects and shoots at targets as though it were some sort of giant.

But Mill's argument can be rescued, and when it is, it leads to the Help Principle. Rather than aggregate utility as Mill did, people can incorporate the concerns of others through empathy. And since we experience empathy with individuals, not with an aggregate[9] – that is, 'one-to-one empathy rather than empathy for a group or for the greater good' (D'Souza and Adams 2016, 124) – we are obliged to maximize happiness within a pair. As with Rawls, this leads to the Help Principle: 'Help someone if your help is worth more to them than it is to you'.

This is not utilitarianism, as becomes clear when three or more people empathize with each other and apply the Help Principle within their group. Instead of maximizing aggregate utility, people in a mutually supporting community, such as a tight military unit, will choose whichever outcome benefits an individual the most.[10] Only when they are not actually reciprocating help to each other, which is likely on a mass scale, will the Help Principle steer closer to the utilitarian option of maximizing total welfare (King 2008, 158–161). This result establishes a clear place for rights, recognizing the importance of individuals. Also, it quells another common critique of utilitarianism, since it means things of value can no longer be completely dissociated from the people who enjoy them (Ibid.).

Deriving the Help Principle this way, as with the derivation from adjusting Rawls' approach, leads to a maxim which is not utilitarian but has been suitably described as 'quasi-utilitarian' – and it corrects the seven commonplace flaws in the traditional concept of utilitarianism. Utilitarianism has been disparaged because it can be self-defeating; ignores past events; doesn't discriminate fairly between people; downgrades promises, fairness, and truth-telling; makes morality a 'slave to the whims of the crowd' (Brett 2014) and doesn't offer any clear rules (Williams and Smart 1973, 91–92). The Help Principle, and its associated form of quasi-utilitarianism, by contrast, corrects for all these oft-cited flaws in utilitarianism. It 'highlights the issues of justice and human rights' (D'Souza and Adams 2016, 124), allows for a 'credible rule that tells us when to lie' (Unnamed Author 2008), and rejects the whims of the crowd when those whims are racist. It even allows promises to have intrinsic value by removing the future bias from utilitarianism (Segal 2015).

The Help Principle, though, should not be thought of solely as a correction of utilitarianism, or of Rawls; it is also an expression of a virtue: empathy (Reiss *et al.* 2015, 262). To act on the Help Principle requires being empathetic towards other individuals, and treating their concerns as if they were one's own. The case for choosing empathy over other virtues as the basis for good character is that empathy is the foundation for deliberate social coordination (Simmons 2014, 97–111). Other important virtues, such as honour, bravery, trustworthiness, and so on, are either not essential for social interaction, or derivative of empathy (Simmons 2014), and thus are excluded by the minimalist criteria. Empathy, by contrast, is essential (Ibid.).[11]

The Help Principle is also a deontological precept: it sets out a rule which instructs action. The Help Principle provides not just the most basic instruction on how one person should interact with another; it offers a well-spring to many other action-centric rules, and an extensive deontological ethical system for a 'myriad of situations' (Brett 2014), with substantial advantages over Kant's famous categorical imperative.[12]

The Help Principle, therefore, leads to a 'quasi-utilitarian system compatible with consequence-, virtue-, and act-based accounts of ethics' (Vardy 2012, 116). It is a system of ethics with a credible basis, which offers coherent advice, and which can provide comprehensive answers to the broad purview of ethical dilemmas.

When humans engage with the world, there is a continuum from their character, to their motives for a given situation, through to their choices and actions, leading to the outcome they bring about, and further consequences beyond those. The most well-established ethical theories achieve coherence and comprehensiveness by focussing on just one part of this continuum: virtue-theorists look to character; Kantians look solely to the action; and Utilitarians focus purely on consequences. Quasi-utilitarianism advocates that we engage with whichever part of this continuum presents itself most clearly. So, when choices about outcomes are clear and information is abundant, consequence-based thinking will be appropriate. When outcomes are indeterminate, but actions can be categorized in certain ways, then rules-based thinking should predominate. And for wicked problems, when the fog of war is most dense, where information about the environment is scarce, and where complexity makes it hard to follow guidelines on how to operate, then a virtue-based, agent-centred approach will lead (King 2023).

A professional ethicist, when tackling a complicated problem, may well draw on principles from all three types of approach. They may advise someone to assess the likely outcome of their actions in so far as it is possible to do so, for example by recommending the most efficient approach; they may also advocate norms and rules, perhaps relating to consent, respecting promises, and adhering to situation-specific codes of conduct; and they would emphasize virtues, such as self-respect, autonomy, prudence, and so forth.[13] But reconciling this smorgasbord of ethical considerations is hard and can overwhelm an individual who wants to behave ethically. It is simply not practical for a soldier in the field. In contrast, the quasi-utilitarian approach based on the Help Principle allows someone faced by a dilemma to assess diverse ethical problems in a single system: it empowers an agent to 'unite (multiple) theories into a new first-order normative theory' (Corrêa and De Oliveira 2020, 793).

Applying the Help Principle and Quasi-Utilitarianism to Ethical Problems in War

So, what does it advise for the six military quandaries set out at the start of this chapter?

The first was Major Howard, ordered to take a bridge held by the enemy. Provided with accurate and reliable intelligence, he faces a situation in which there is little uncertainty between his actions and the outcomes they will generate. Choices in these cases are relatively straightforward and well-defined; Howard may even be able to set out a menu of possible courses and choose between them, confident that nothing is missing from the list.

In a paradigm case of this sort, Major Howard will know not just how many enemy fighters are in the area, but how they will react to an attack, their propensity to run or fight, their weapons and so on. He could generate a menu of options, including 'use stun and smoke grenades, and allow them to run away', 'kill them all', and so on.

In this scenario, the closed nature of the problem, and the availability of decent information, means Howard can identify which courses of action will take the bridge and then select from that sub-category the option which serves secondary goals best. The quasi-utilitarian approach will advocate choosing a course of action which minimizes the number of life-years lost by his unit, since this option would be preferable to any other option resulting in more death.[14]

Killing enemies to take the bridge requires a higher purpose: a reason to justify the causing of harm, since this departs automatically from maximizing all-time value. Traditionally, the six principles of a just war have set out what a higher purpose might be[15]: there must be a just cause, the right intent, a reasonable chance of success, peaceful avenues must have been exhausted, the war must be conducted proportionately, and it must be initiated by a legitimate authority. These principles need to be adapted to modern times, when the distinction between war and peace is blurred, when there can be wars within wars, and when legitimacy can span several competing authorities in complex ways. But when the consequences of violence are absolutely certain, as they are in Major Howard's hypothetical case, a calculation can be made, and some legal norms become superfluous. The Kosovo war, for example, was not entirely lawful,[16] but it was certainly a good war, because genocide was prevented. With perfect foresight, Howard can come to similarly firm conclusions with minimal direct reference to rules.

Should the major put himself in jeopardy to save one of his men? The reciprocity rule limits how much risk someone should accept to that which the people they are rescuing would themselves be prepared to bear.[17] Howard should not sacrifice himself for a single other individual, but it would be right for him to give up his life to save several – and, in theory, with perfect information, it would be possible to calculate exactly how many (King 2008, 166–167).

Now consider the situation faced by Police Commander Trenton, responsible for fighting terrorists: opponents who believe they are at war, even though Trenton and her government regard the terrorists merely as criminals.

Trenton faces much more uncertainty than Major Howard. She does not know all the potential terrorists, or how they will behave (see Grint 2010, 306–313).[18] Also, she is not in a position to describe all the possible consequences; the outcomes she can generate cannot be listed on a menu. But Commander Trenton's situation does have the advantage that the immediate effects of her actions will be known.

Trenton is in the domain of norms: she should set out rules of behaviour (Ibid.). She will want to establish a means to prioritize terrorism alerts, so she can send more resources to the most dangerous areas. She will want her anti-terrorism officers to employ safety protocols, and have standard procedures for dealing with potential incidents, such as 'save innocent people before trying to arrest terrorists'. Some of these norms or rules may become internalized – embedded in the instinctive thinking of anti-terror officers so that they do not require active thought. This would be appropriate: it will save both time and mental energy and thus allow for split-second decisions (Weyman and Barnett 2016, 131–139). Given the level of uncertainty, the time pressure created by some terror alerts, and the unavailability of key information, acting on rules and protocols in this way is much better than trying to operate as if outcomes could be listed and compared – Major Howard could do that, but Commander Trenton can't.

There will be occasions when the rules set by Trenton are inappropriate, but trying to determine which situations they are is an unpromising enterprise. Further, trying to seek exceptions undermines the general application of the rules themselves.[19] Hence, breaking rules will usually need to incur a sanction, even in cases where no negative consequences follow, such as 'victimless crimes'.

Again, in this instance and others, quasi-utilitarianism makes recommendations which differ significantly from those of an act- or rule-utilitarian, or a Kantian approach. Quasi-utilitarianism advises an individual anti-terrorism officer presented with a situation where breaking a rule may bring about a better outcome to decide on the presumption that the breach *will* be exposed, with all that entails, whether it actually is exposed or not (King 2008, 148). And Police Commander Trenton should update her rules and protocols, but only occasionally: when it emerges they are based on out-of-date patterns in how the world works, or when it's clear their impact diverges substantially from the Help Principle (Ibid., 194).

Finally, note also that the advice for Commander Trenton departs from the traditional just war theory. Few would consider the terrorists she is trying to beat a 'legitimate authority', which implies countering terrorism requires a different approach to war-fighting. But quasi-utilitarianism suggests this distinction is artificial: legitimacy varies by degrees; it is not an all-or-nothing concept. Like the distinction between war and peace, the medieval notion that only some actors may use violence to seek a political effect is outdated. Our ethical advice needs to be up-to-date, too.[20] If Trenton is to use the fact her terrorist opponents are not legitimate to harm them, rather than merely to neutralize their efforts, she will need a higher justification – such as to prevent them from harming others.[21] The illegitimacy of the terrorists may be a factor in this, but it is not, in itself, a reason to inflict harm.

The third example was the problem faced by the Academy Commandant, Brigadier James. He has even less information than Major Howard and Police Commander Trenton: he cannot know what problems will be faced by the young women and men he's training when they reach their prime military years, probably a quarter of a century away. If he fails badly, his country will be without an effective fighting force. His situation is 'more complex, rather than just complicated – that is, (it) cannot be removed from its environment, solved, and returned without affecting the environment' (Grint 2008).[22]

Although the military training portfolio is daunting, the Brigadier will have more agency in how he tackles his situation than either Major Howard or Police Commander Trenton, whose problems were confined within tight limits. James' situation offers much greater freedom of action; the ambient uncertainty empowers him to shape events.[23]

Brigadier James cannot set down a menu of possible outcomes: too much is unknown. Even rules and norms are a poor guide, since they depend on regularities which are unlikely to persist into the world for which he is preparing his soldiers, several years away.

The only thing which will be constant for the future leaders being trained by the Brigadier will be themselves, which means having the right characteristics comes to the fore. Even though he knows little about the environment his trainees will face, the Brigadier can instil certain virtues in his troops. In social situations, quasi-utilitarianism suggests empathy for their fellow soldier will be the principal trait he should promote. By rating the concerns of others as if they were their own,

James' cadets will become more able to navigate the complex problems they will face; other martial virtues, such as bravery, a willingness to take the initiative, and leadership, all turn on empathy in one way or another.

In the years after their teaching from Brigadier James' academy, the next generation of military leaders will face not one decision, but many. Their options will come in clusters and there will be much path dependency; they will face a more-or-less continual process of deciding which route to take as they promote through the ranks – often they will need to define the way themselves. They should continually re-assess whether rules or consequences can alleviate any part of the situations they face. By making these choices empathetically, they are more likely to choose a better path in the years ahead.

When to Break Rules

Quasi-utilitarianism suggests rules, norms, and laws should generally be respected, but sometimes broken and occasionally dismissed completely (King 2008, 192). The benefits brought by rules (they simplify decision-making, guide activity close to the quasi-utilitarian outcome, provide advice when consequences are unknowable, coordinate different people's activity, and clarify personal responsibility)[24] can only ever be outweighed by harms when outcomes are at least partially knowable. To be justified, an incident of rule-breaking must bring a better outcome even if it is exposed (whether it actually is discovered or not)[25]; hence, Admiral Nelson was right to put a blind eye to the telescope so he could ignore his orders. This formula accounts for rules of different weight and importance, factors in the consequences of the breach, and allows for a separate decision on whether the rule breach should be kept secret. It also differs substantially from rule utilitarianism.[26]

This formula helps resolve the problem faced by Sven, the fourth example at the beginning of this chapter. Captured during the Bosnian war of the early 1990s, Sven is held at gunpoint, handed a revolver, and ordered to kill one of his fellow prisoners. If he does not, then his captors will kill ten POWs. Sven soon realizes the dilemma is straightforward; there is no realistic prospect of turning the revolver on his captors, escaping, or raising the alarm.[27] Sven's calculus is not as simple as some would believe – his decision process takes in more than just saving lives, since Sven, as a medic, has adopted the Hippocratic oath and takes very seriously the rule that he should not kill.

What would Sven do, if he takes quasi-utilitarian advice? If the scenario is exactly as stated, then Sven should apply the rule about rules: he should kill one POW to save nine, because breaking his commitment to not kill, even when discovered, and even when factoring his self-loathing for the deed and the undoing of his virtue, brings about a much better outcome than allowing all ten to die. Note that Sven should also check the facts, and seek out some other way to break the dilemma – perhaps committing to bring his captors to justice after the war. He is still obliged to reconcile the rules he follows with good outcomes whenever he can.

The quasi-utilitarian approach does not simply mean outcome-based thinking trumps rule-based thinking, or that virtues are ignored. Rather, which of these ethical elements features in the advice will depend on what information is available. What matters most is the point in the decision-making process – the continuum from character to action and consequences – which is in focus, and it will vary according to the scenario.

Finally, we come to the fifth and sixth dilemmas, those faced by Private Wills and Captain Ivan, both of whom are given orders which make them uneasy, but in different ways. For Wills, the experience of battle over 10 weeks has created a revulsion to fighting. Captain Ivan, meanwhile, doubts the wisdom of his superiors.

Private Wills is experiencing a natural reaction to death at close quarters – a counterpoint to the enthusiasm for war he felt a few weeks earlier. Was he right to like fighting then, or is he right to hate it now?[28] These contrasting sentiments evolved in humans through many generations of trial and error and helped our ancestors survive to reproduce. Only some of these instinctive feelings remain appropriate for decisions people face today. Like two overlapping circles slowly moving apart, as modernity separates from its past, the area of intersection is gradually shrinking. The challenge, when instinctive norms depart from the right course of action – such as when Sven must kill to save nine prisoners of war – is to determine whether it is an isolated aberration or evidence of a longer term trend for which our ethical instincts are ill-adapted. Whether we need to revise our norms depends on how often we are faced with scenarios like Sven's – long-term climate change calls for systematic behavioural change which a single hot summer's day does not. So, for Private Wills, we can judge he is no longer effective in the fight. His seniors should keep him back, and if they do not, then it is best to keep his head down. Whether it is right for soldiers like Wills to be encouraged to feel bloodlust, or a reluctance to kill, will depend on the broader circumstance his unit is facing.

We need to respect our moral intuitions, while also seeking moral progress, to bring those intuitions in line with contemporary needs. We need to adjust a moral intuition or norm when doing so regularly brings about a better outcome, while recognizing there may be elements of the old norm which can and should still carry weight because we cannot be sure what awaits us in the future. This is not an easy calculation, but note the vagueness lies not in the moral system, but in our incomplete understanding of the world – it is principally about understanding facts rather than values. Furthermore, it is perilous to change all our instincts at once – to use the famous analogy of Otto Neurath, who drew parallels with sailors trying to refit their boat at sea, we must stand on some timbers if we are to refresh the rest (Neurath 1973 [1921], 158); we should reshape our moral instincts 'only by gradual reconstruction' (Ibid.).

Captain Ivan's situation, instructed to fire his artillery 20 miles west to a location he believes is the site of a school, presents a different challenge to military obedience. Ivan exists within a 'community of practice' with localized moral codes (Kaurin 2020, 96–118): there is a strong presumption that a senior officer has good reason to issue a seemingly flawed order; even when challenge and dissent are appropriate, they should only rarely entail disobedience.

Ivan needs to relay his concerns up the chain of command: are they sure there's no school there? Suppose he receives a satisfactory response – for example, the school has become a headquarters for enemy troops, can he trust his superiors on these facts? If he can, then he should fire; Captain Ivan should generally obey the orders he is given in this case.

But if the facts are still in doubt – perhaps his seniors have a poor record – and there is no further avenue to clarify the outcomes he may bring about, he should defer to rules. There are two relevant rules for him here: 'obey orders' and 'don't commit war crimes'. The second of these rules prevails over the first, so he shouldn't fire as directed. And if Ivan's refusal means someone else follows the dubious order instead, then he should hide his disobedience by aiming his artillery rounds at a different spot.

Ivan's obedience is conditional: his superiors must treat him in certain ways for them to retain their authority over him. They should not direct him to break international law, and they should give him the appropriate kit, keep him informed on certain things, pay him, and so on.

The captain may still not want to fire his shells, even when he accepts the school building is now an enemy HQ – perhaps because he doesn't want to cause harm, or because he is uneasy about the overall war effort. But the full consequences of not doing an action need to be considered, in particular on whether it empowers someone else to do even worse. From time to time, people have to do harmful things: in those cases when the behaviour of others means any better actions would lead to a worse outcome. Whenever these situations arise, we are required to try to remedy the underlying problem. Only when something greater is at stake should we jettison our personal integrity by committing a bad deed; and in place of that integrity, we must confront the root cause which compels us to pursue the good outcome through harmful actions. Ultimately, Captain Ivan fires his artillery – whether to hit the enemy HQ or to miss the school – to make the world less hospitable to moral dilemmas.

Conclusion

In order to answer what ethical military action involves, this chapter explored the basis of ethics, settling on a quasi-realist explanation. It then established the Help Principle and the associated methodology of quasi-utilitarianism as the most suitable guide to ethical behaviour and explained why it was better than other systems of ethics.

Advice from quasi-utilitarianism will vary according to the situation faced: being ethical requires us to think and act in different ways according to the character of the scenario which confronts us. Different parts of this chapter have shown how this theory of ethics, at once novel and hybrid, is superior to each of the standard approaches which always root ethics in just one place – outcomes, rules or virtues. Finally, equipped with this explanation of what is ethical behaviour, this chapter then applied it to military affairs, and so set out advice for what to do in war – advice which is credible, comprehensive, and coherent.

A summary of what people should do is this:

> Where outcomes are known or easily knowable, choose the best outcome available.

1 The best outcome will be the one which maximizes all-time value. Only direct welfare counts – not the happiness people derive from how others fare. Past promises should be given weight as if they were current.[29]
2 When people reciprocate help to each other, generally in smaller groups, the best outcome will tend to be the one which helps an individual the most; in larger groups with less reciprocal behaviour, it will tend towards maximizing overall welfare.
3 In a conflict, actively discounting the welfare of others requires a higher justification – a reason beyond the immediate situation itself; usually to prevent harm on a much greater scale.
4 For a military unit with reliable information on what can happen and a higher justification to fight, resolve the trade-off between preserving the welfare of the unit's members and risking that welfare to achieve military goals by listing and then comparing the set of possible outcomes.

When outcomes are unknowable, or they contain too much uncertainty to guide behaviour, then rules become the central determinant of what is ethical. Rules can encapsulate knowledge, simplify decision-making, coordinate different actors, ascribe responsibility, and instil certainty where it would otherwise be lacking. Where a community is working well, we should generally respect the rules on which it runs.

1 Rules can be established for a variety of specific activities, so as to maximize the all-time value within that activity. Quasi-utilitarianism offers many rules (Brett 2014) and sets of rules can be developed for most activities, including warfare.
2 The rules for warfare should, as a whole, direct individuals and groups to act so as to minimize all-time harm caused. To make deterrence credible, and in line with the reciprocity rule, acts of violence will often warrant a violent response.
3 Rules should only be broken when doing so is likely to bring about a significantly better outcome, factoring in the harm caused by breaking the rule as if it were discovered, whether the breach actually is discovered or not. Consider whether to break the rule secretly as a separate decision, according to whether you have a reason to believe people will not behave appropriately if informed about the breach.

4 A rule, or system of rules, should only need to be changed occasionally: when it emerges that they are based on out-of-date facts about the world, or when it's clear that their impact diverges substantially from the Help Principle. Warfare accelerates the rate at which rules need to adapt to changing circumstances.

Finally, when even rules offer little or no valuable guidance, we are forced to rely on virtues. When acting alone, these will simply be the manifestation of our character; when we act with others, the principal virtue is empathy: we should act with empathy towards others. Empathy enables people to understand other people's problems as if they were their own; it allows solutions to be imagined. Empathy helps a military unit bond and fight together, and empathy towards victims of conflict can guide how they are treated. Empathy towards the enemy will reflect the empathy they show themselves.

At each point, an agent is concentrating on the part of the decision-making continuum most in focus. When this offers more than one option, since the ultimate goal of our interactions with the world is to affect it, the face closest to that effect will be most important: we should not slavishly stick to our virtues when we are sure they will lead to bad actions, nor dutifully stick to a rule when we are certain it will cause a negative outcome overall. But neither should we abandon our virtues, rules or instincts without regard: they carry weight, and it is worth maintaining respect for them, even at the cost of moderately bad consequences from time to time.

Our general approach should be between pacifism and militarism. We should be neither allergic to war nor enthusiastic about it. Deliberate violence is bad because it causes harm, but there are occasions when causing harm is the least bad thing to do. Fighting is sometimes the best response to a situation, and going in hard can be better than minimizing damage.[30]

Also, there is a strong basis for self-defence. This emerges through the principle of reciprocity. As we empathize with people to the degree that those people empathize with themselves, we have a general case for reciprocating violence – inflicting unwanted pain is the inverse of empathy. Empathy for the victims of violence means, in turn, we have a moral case, often a duty, for reciprocating on their behalf – that is, a responsibility to protect.

And: scale matters. When a militant danger threatens only minor damage – a challenge, perhaps, to the life of an individual, or the possibility of injury to several – the right response will be of a quantitatively different sort to a danger at a mass scale, which threatens whole communities, nations, or humanity itself. Rights and norms are important, and they are worth making great sacrifices for, but we should not sacrifice the world for them.

With all ethical decisions, the way to make choices about war and in war changes according to how much reliable information is available. It is exceptionally rare to know the full consequences of our choices; and, often when we think we do, we are mistaken. So, there must be norms: rules which govern our behaviour, derived straight from the steps shown earlier in this chapter. The norms of war should be those which, when generally followed, lead to the best outcomes.

But setting rules for war fighters differs fundamentally from setting rules in most other affairs, because war-fighting is fundamentally about causing harm. Hence, the rules which govern fighting, and causing violence in general, require a higher purpose for any sort of violence to be legitimate. The antipathy for an enemy which is so central to war can only be justified when it flows from an empathy towards others.

Next time the alarm blares out across the base in Baghdad, perhaps with real incoming fire this time, our empathies should be with those running for cover. We do well to hope those firing the mortars and rockets into the camp have empathy for us, too.

Notes

1 For example, the 'War of Jenkins' Ear', fought to avenge the removal of Captain Jenkin's outer ear some 8 years before the fighting began.
2 This approach was first described as 'quasi-utilitarian' by Charlotte Vardy and Peter Vardy, *Ethics Matters* (SCM Press, 2012), 116. It is an appropriate description.
3 Arbitrary in the sense that there is nothing intrinsically good or bad about, say, climate, the patterns of natural disasters, or the pace of agricultural innovations, which favoured some people and their moral codes over multiple generations.
4 A novel and persuasive argument has been made that the things which ethics are most like is sub-atomic particles. See Iain King and Myles King, 'Ethical Truth in Light of Quantum Mechanics.' *Philosophy Now* 156, June/July 2023.
5 'Ethicists as diverse as Kant, Mill, and Ross have assumed that an adequate moral theory should not allow for the possibility of genuine moral dilemmas' (McConnell 2018).
6 Almost all virtue-based systems of ethics advocate a blend of good characteristics rather than a single one, according to KJ Shanahan and Hyman (2003, 197–208).
7 As noted by two moral psychologists,

> perhaps it is possible to live a moral life guided by the categorical imperative (or) by the principle of utility . . . (but) in order to do this, it is necessary to know which sensibilities we should . . . suppress, moderate, modify, or redirect. Flanagan and Oskenberg-Rorty (1993, 1)

8 Kant himself, and some of his supporters, argue this critique is unfair. See Rickman (2011, 10–12).
9 A counterargument to this may suggest that we can empathize with a group, for example, a platoon which surrenders together. This chapter asserts that this either involves empathizing with individuals in the platoon, one-by-one; or it is a different concept of empathy not relevant to this point.
10 For details on how this result emerges, including how the notion of 'welfare' here differs from the standard utilitarian conception, see King (2008, 158–161).
11 The pre-eminence of empathy is reinforced by a further argument, that it is the only virtue with attributes which fit the attributes expected of a virtue 'while its opposite does not'. See King (2008, 74).
12 Kant's maxim ('Act only in accordance with that maxim through which you can at the same time will that it become a universal law') cannot accommodate actions which are unsustainable if all people do them. Hence, Kant would deem it immoral to arrange surprise birthday parties. The Help Principle avoids this flaw.
13 Some institutions already incorporate all three approaches: a planning committee, for example, will primarily judge based on the outcome proposed, while being sure to

follow certain rules which ensure fair competition, and maintaining certain virtues, such as probity.

14 A more sophisticated calculation would factor in injury and other harm, and may draw on the concept of 'Quality Adjust Life-Years'.

15 Principles about the conduct of war are also relevant but are usually deemed to come into play only when there is a fundamental justification for politically motivated organized violence.

16 The primary conclusion of the authoritative Independent International Commission on Kosovo was that the 1999 war was moral but not entirely lawful. See King and Mason (2006).

17 Although there are significant differences, the reciprocity rule is similar to the Golden Rule: 'Apply help to others as they would apply it themselves'. See King (2008, 109–114).

18 Trenton is dealing with what Grint labels a 'tame problem'.

19 Since rules usually ascribe responsibility to an individual, trying to deduce the contribution of someone's action to an amorphous or indeterminate outcome also runs against providing clear ethical guidance.

20 Determining which people or institutions are a 'legitimate authority' is a norm; it sits alongside other norms, not above them.

21 To proactively harm terrorists, Trenton would need more knowledge about likely outcomes than is offered in this scenario.

22 Brigadier James faces what Grint (2008) labels a 'wicked problem'.

23 It is a truism in decision-making that freedom of action declines as information increases, partly because the new information forecloses some options, and partly because others will make choices that constrain your own.

24 Note that personal responsibility can be established through outcome measures, too – such as assigning objectives to individuals.

25 King (2008, 148–149): 'Deceive only if it will change behaviour in a way worth more than the cost of lost trust, were the deception discovered'.

26 Rule utilitarianism traditionally seeks to align rules with producing the best outcome; by several compelling accounts, it does not give intrinsic moral weight to rules other than as a heuristic.

27 There are several documented cases of captives being forced to conduct horrendous acts during the Bosnian war; this scenario is similar to 'Jim and the Indians', posed by Bernard Williams. See Williams and Smart (1973, 97–99).

28 Drawn from King and Perrotta (2014, Chapter 1).

29 Traditional utilitarians look only to the future costs and benefits of breaking a promise, and would say giving weight to past behaviour made on the basis of a promise is the 'sunk costs fallacy'. This degrades commitments and undermines military covenants. Promises can be broken occasionally, but only when the benefits of doing so exceed the full cost already invested in that promise. For more on which commitments should be made, and when they can be broken, see King (2008, 140–144).

30 It should be noted that, compared to their subordinates, military leaders will more often face wicked problems. These require a focus on virtues and, collectively, the culture of the institution they lead. Although a fighting culture based around a will to harm may seem at odds with the primary virtue of empathy, this is not automatic, for example because it can deter violence by others. For more on this, see King (2023).

References

Blackburn, Simon. 1998. *Ruling Passions*. Oxford University Press.
Brett, Chandler. 2014. *24 and Philosophy*. Blackwell Philosophy.

Corrêa, Nicholas, and Nythamar De Oliveira. 2020. 'Metanormativity: Solving Questions about Moral and Empirical Uncertainty.' *International Journal for Moral Philosophy* 19(3), 793.

D'Souza, Jeevan F., and C. Kelly Adams. 2016. 'On Measuring the Moral Value of Action.' *Frontiers of Philosophy in China* 21(1), 22–136.

Fisher, Andrew. 2014. *Metaethics: An Introduction.* Routledge.

Flanagan, Owen, and Amelie Oskenberg-Rorty. 1993. *Identity, Character and Morality: Essays in Moral Psychology.* MIT Press.

Grayling, AC. 2019. *The History of Philosophy.* Penguin.

Grint, Keith. 2008. *Wicked Problems and Clumsy Solutions.* BAAM Publications.

Grint, Keith. 2010. 'The Cuckoo Clock Syndrome: Addicted to Command, Allergic to Leadership.' *European Management Journal* 28(4), 310–313.

Haist, Matthew. 2008. 'Deterrence in a Sea of Just Deserts: Are Utilitarian Goals Achievable in a World of Limiting Retributivism.' *Journal of Criminal Law and Criminology* 99(3), 789.

Joyce, Richard. 2008. 'Precis of the Evolution of Morality.' *Philosophy and Phenomenological Research* 77(1), 214.

Kaurin, Pauline Shanks. 2020. *On Obedience: Contrasting Philosophies for the Military, Citizenry and Community.* Naval Institute Press.

King, Iain. 2008. *How to Make Good Decisions and Be Right All the Time.* Continuum.

King, Iain. 2014. 'Moral Laws of the Jungle.' *Philosophy Now* 100, 20–22.

King, Iain. 2023. 'What Is Ethical Leadership?' *King's Centre for Military Ethics.*

King, Iain, and Whit Mason. 2006. *Peace at Any Price: How the World Failed Kosovo.* Cornell University Press/Hurst Publishing.

King, Iain, and Lou Perrotta. 2014. *Making Peace in War.* Amazon Publishing.

King, Iain, and King, Myles. 2023. 'Ethical Truth in Light of Quantum Mechanics.' *Philosophy Now* 156(June/July).

Korsgaard, Christine. 1985. 'Kant's Formula of Universal Law.' *Pacific Philosophical Quarterly* 66(1), 24.

Louden, Robert B. 1984. 'On Some Vices of Virtue Ethics.' *American Philosophical Quarterly* 21(3), 229.

McConnell, Terrance. 2018. 'Moral Dilemmas' in Edward N. Zalta and Uri Nodelman (eds) *The Stanford Encyclopedia of Philosophy.* Metaphysics Research Lab, Stanford University.

McCormick, John P. 1998. 'What's Wrong with Liberalism? A Radical Critique of Liberal Political Philosophy.' *Political Science Quarterly* 13(4), 748–750.

Mill, John Stuart. 1991 (1863). *Utilitarianism.* Oxford University Press.

Neurath, Otto. 1921. 'Anti-Spengler.' *Empiricism and Sociology* 1973(1921), 158.

Rawls, John. 1971. *A Theory of Justice.* Harvard University Press.

Reiss, Annie, Yigal Shafran, and Esther-Lee Marcus. 2015. 'Bioethical Dilemmas in Jewish Thought on Medical Resource Allocation: The Coexistence of Opposing Views.' *Medicine and Law* 34, 251–262.

Rickman, Peter. 2011. 'Having Trouble with Kant?' *Philosophy Now* 86, 10–12.

Rosen, Frederick. 2005. *Classical Utilitarianism from Hume to Mill.* Routledge.

Rowland, Stuart, and Tracey Rowland. 1995. 'The 'Political Values' of the "Public Conception" in the Work of John Rawls.' *University of Queensland Law Journal* (18), 345.

Sartre, Jean-Paul. 2007 (1946). *Existentialism Is a Humanism.* Yale University Press.

Sayre-McCord, Geoff. 2012. 'Metaethics' in Edward N. Zalta and Uri Nodelman (eds) *The Stanford Encyclopedia of Philosophy.* Metaphysics Research Lab, Stanford University.

Segal, Robert. 2015. *Vocabulary for the Study of Religion*. Brill.

Seth, James. 1908. 'The Alleged Fallacies in Mill's "Utilitarianism".' *The Philosophical Review* 17(5), 469–488.

Shanahan, K.J. and HR Hyman. 2003. 'The Development of a Virtue Ethics Scale.' *Journal of Business Ethics* 42, 197–208.

Simmons, Aaron. 2014. 'In Defense of the Moral Significance of Empathy.' *Ethical Theory and Moral Practice* 17(1), 97–111.

Unnamed Author. 2008. 'How to Make Good Decisions and Be Right All the Time.' Publishers' Weekly. www.publishersweekly.com/978-1-84706-347-2

Vardy, Charlotte, and Peter Vardy. 2012. *Ethics Matters*, 116. SCM Press.

Weyman, Andrew, and Julie Barnett. 2016. *Heuristics and Biases in Decision Making about Risk*. Routledge.

Wielenberg, Erik J. 2016. 'Ethics and Evolutionary Theory.' *Analysis* 76(4), 502–515.

Williams, Bernard, and J.J.C. Smart. 1973, *A Critique of Utilitarianism*. Cambridge University Press.

5

A NATURAL LAW BASIS FOR PRACTICAL MILITARY ETHICS

Rufus Black

Having a sound basis for military ethics is as important as it has ever been. Doing what is right has always been intrinsically important, but it is also clearer than ever that acting ethically confers strategic advantage in building alliances, maintaining the political commitment to see a war through, and securing the support of people being defended. In addition, it strengthens organizational cultures, and it protects individuals from moral injury. For any modern military, these are all critical needs. Conversely, as history, both recent and longstanding, has taught us, unethical acts, especially war crimes, are seriously injurious to all these causes.

The Tasks of Military Ethics

To meet these needs, military ethics needs to be both intellectually sound and practically deployable in a military context. We will come to philosophical rigour shortly, but we will start with what it needs to do practically:

- Enable military personnel engaged in armed conflict to make very rapid, reliable, and consistent ethical decisions under conditions of extreme pressure.
- Enable military and other personnel to make routine ethical decisions in situations outside of armed conflict easily, without having to think them through every time.
- Provide the thinking tools to enable the ethically complex and grey issues of military life to be carefully analysed and clearly explained.
- Provide an account of authority that explains when it is ethical to follow orders, including the basic order to go to war, even when there are doubts about whether the order is ethical and when it is right to disobey those orders.

DOI: 10.4324/9781003312925-6

- Align the behaviour of service personnel with the legal constructs that govern the conduct of armed conflict.
- Help shape military culture and character of military personnel to create consistent ethical behaviour at all levels of the organization and across the military's full range of activities.
- Provide clarity on how to understand ethically the sort of 'high stakes events' that are associated with causing moral injury, like being involved in the killing of innocent civilians, which could, depending on what happened and how those events are ethically understood, involve the 'transgression of one's deeply held moral convictions or beliefs about right and wrong' (Phelps *et al.* 2022, 1; see also Litz *et al.* 2009, 695–706). At a minimum, a sound ethical approach will not leave military personnel with 'ethically insoluble dilemmas' especially involving killing, which seem particularly associated with moral injury, and ensure that they don't feel moral guilt about actions, which correctly understood might have involved tragic losses but were not wrong (Phelps *et al.* 2022, 2).

For a theory to fulfil these practical tasks of military ethics, it needs four features.

First, it needs to be both a theory that provides principles and rules to analyse complex ethical problems and guide decisions and a virtue theory that recognizes the character-forming significance of decisions and that often people's decisions will come from people acting consistently with their character, rather than detailed ethical analysis. In doing so, we will need to challenge the often-held assumption that ethics of duty (deontological ethics), which provides a rich set of principles and rules, and virtue ethics belong in different logical families but rather they are integrally related elements of a single theory.

Second, the theory needs to be capable of providing simple consistent rules that can be used easily in high-pressure situations but have the sophistication to reveal, when needed, the texture of complex ethical problems. It is especially important that the theory has the analytic resources to deal with complex ethical issues, which arise in military situations including well-intentioned actions that have significant harmful side effects; participation in wrongdoing; maintaining or conversely breaching promises, including commitments to maintain information as private or confidential; and the handling of communications when the straightforward sharing of information may have serious negative consequences. Clarity about complex ethical matters is not just important to ensure military personnel do what is right but also to reduce the risk of unnecessary moral injury.

Third, the theory needs to have as an integral part of it an account of the ethics of authority because command structures are a fundamental feature of military organizations and operations. It needs to be able to explain why it is ethical to follow orders even when you don't agree with them and, equally, when it is ethically required to disobey them. In tackling these questions of authority up to and including political authority, a theory that underpins military ethics must have the scope to tackle related questions of jurisprudence and political philosophy.

Fourth, it needs to be consistent with the underlying ethical logic that governs the laws of armed conflict and international law more broadly. This, of course, assumes that these laws are themselves ethically sound as this chapter will maintain. Part of the importance of this requirement in a military context is that an ethical framework that justified even in extreme cases a soldier committing war crimes would be very problematic for the military forces of a nation that is governed by and respects those laws. Similarly, because of the centrality of the military operating with the rule of law domestically, an ethical theory that is consistent with the broad premises of domestic law is important.

Natural Law Theory as a Foundation for Military Ethics

For a long time in Western militaries the ethical theory that has been used implicitly and sometimes explicitly has been natural law theory. We will argue that is also a theory that fulfils these four requirements for practical military ethics. That is why in this chapter we will consider it carefully as a continuing foundation for military ethics.

While we don't often see natural law theory explicitly named as a form of military ethics, there are very few discussions of military ethics that don't recognize just war theory as central to ethical deliberations about the use of force. Similarly, discussions of what is ethically right and wrong are almost always shaped by consideration of the laws for armed conflict. The origins of both just war theory and the laws of armed conflict are in natural law theory.

Nowhere is that relationship clearer than in the famous work by Hugo Grotius *On the Rights of War and Peace* (1625),[1] which brought together in a brilliant and enduringly influential synthesis the ethical foundation of the customary international law that governed matters of war and other aspects of international law more broadly.

Grotius' work to lay those foundations begins in his first chapter with his exposition of the natural law theory. Central to his explanation about natural law theory is his statement under the subheading 'The Law of Nature Defined' in which he says 'Natural Right is the Rule and Dictate of Right Reason, shewing the Moral Deformity or Moral Necessity there is any Act' (Grotius 1625 (2005), I.1.10.1 (p. 150)). Critically, he is pointing out that the essence of natural law or natural right theories is that they are about the application of sound reasoning. The 'natural' part stands in defence of the realist qualities of right reason. There is an objectivity to sound practical reason (i.e. the reasoning of ethics). Natural law theory is not as it is sometimes misunderstood or misrepresented about conforming to how things behave in nature (Finnis 1980, 33–48). It certainly does not try to draw an ethical conclusion from an observation about nature. It does not commit the naturalist fallacy of trying to get 'an ought from an is' (Grisez 1987, 99–151, 102). As

Grotius and many who came before and after him have made clear, it is all about sound practical reasoning.

Standing in that tradition is the contemporary natural law account of practical reasoning that we will draw on, which was one developed by a school of philosophers principally led by John Finnis, Germain Grisez, and Joseph Boyle.[2] It is worth noting that this school of thought sometimes gets dismissed or criticized because these philosophers and some who have built on their views, like Nigel Biggar, have defended traditional Catholic views on topics like sexuality and gender or other conservative positions (see Biggar and Black 2000). As I have noted elsewhere, I don't see these authors' conclusions as the logically necessary outcomes of this form of natural law theory and I don't agree with their conclusions. Those who criticize or dismiss natural law theory because of its associations with people who hold particular views risk missing the analytic rigour the theory offers. Any attempt at theoretical or personal 'guilt by association' is an unsound and cheap form of criticism.

Intention

In our case, the practical reasoning of which Grotius speaks begins not with any claims about the natural world with the question, 'what do we intend?'

This question of intention has been a central subject for ethical inquiry in Western law and society. It has proved to be an insightful starting point for understanding right and wrong. Critical to our project is the way it enables us to make important ethical distinctions between acts which otherwise have the same outcomes. For example, the distinction between murder and manslaughter hinges on whether there was an intention to end a person's life or not, just as the distinction in civil law between intentional torts like assault and battery and negligence pivots on whether the harm was intended, or the result of a failure to attend to a duty of care. Through examples like these, we come to understand that the nature of someone's intention fundamentally alters the character of the act and how we understand the ethical significance of the outcome.

The intention we are describing here embraces both 'the end' that we are pursuing and 'the means' we plan to use. We will consider each in turn.

What Is the Good We Intend?

When we inquire into intention, firstly what matters is, what is the end or outcome that someone is pursuing? In thinking about those ends or outcomes, natural law theory builds on Aristotle's insight that what practical reason seeks is human well-being or flourishing.[3] Sometimes that flourishing comes from the pursuit of personal goals like, health, being skilful in work tasks and knowledge or relational ones, like friendship or even societal ones, like justice.[4] In military settings, it is easy to see soldiers act in defence of themselves, their comrades, or in the pursuit of the just outcomes that define their mission.

When people pursue those goals of human well-being, it is important to recognize that those qualities, from friendship to being skilful in work tasks, aren't abstract universals. They always have a culturally embodied form.[5] The character and boundaries of friendship vary from one society to another just as the nature of work tasks varies widely. Even within societies the character of these good ends that are integral to human flourishing varies. Often for example, in a military context, the experience of friendship has a very particular quality to it born of the nature of military life and culture. An ethical theory that can recognize these unique expressions of what we understand to be valuable, while still holding to the fact that there are ethical boundaries, that there are indeed good and bad intentions, provides a rich palate to describe the ethical realities such as what makes for human well-being.

Occasionally, people do intend ends that thwart or damage human flourishing. For example, if they intentionally kill an innocent person, inflict pain out of revenge, or deliberately destroy works of human creativity. This is the most foundational type of practical unreasonableness because it involves contradicting the very purpose of being practically reasonable, which is to pursue what is humanly worthwhile.[6] This sort of unreasonableness usually characterizes the most serious ethical failings as it does the most serious criminal offences. In military settings, war crimes, both ethically and legally, commonly involve sort of deliberate harm and destruction.

The analysis of our intended ends makes an important contribution to the ethical underpinnings of the criteria of just war theory. What it comes to the justice of going to war (*just ad bellum*) the need for a good end explains why we have the criteria that there must be a just cause for going to war in the first place. The end we are seeking is to remedy a serious injustice. And equally, it grounds the requirement of 'right intention'. In other words, the end we intend must be remedying the injustice not some other self-interested end pursued under the guise of righting a wrong.

The focus on ends is important because often those who favour ethical approaches in the broadly utilitarian family of theories at times assert that other theories don't adequately recognize the ethical importance of consequences. Natural law theory focuses on consequences in two important ways. It considers the consequences we intend to bring about and our responsibility for the unintended consequences, which are the side effects of our intended actions.

Natural law theory also offers two important clarifications about our ethical relationship to these different types of consequences. First, as a theory of human practical reasoning, it only asks that people consider consequences that a human could reasonably foresee. Part of why utilitarian theories fail in practice and have a core conceptual problem is that, especially in settings like military conflict, determining all the consequences that might flow from a decision is a vastly complex task far beyond what any person can do, let alone in the circumstances of a time-pressured decision.

Second, as we saw earlier, the ethical character of the consequences we cause can be fundamentally coloured by what we intended. An innocent person's death that is the result of a well-intentioned act that goes wrong is a tragedy but the same death that is intended is a crime. Without the ability to consider intention many of the important ethical distinctions that provide insight, like the distinction between tragedy and wrongdoing, can't be made. In the context of military conflict, where much harm occurs, there is a real importance to be able to make these kinds of distinctions for both ethical clarity and psychological well-being. It is a very different psychological journey for someone coming to terms with their human limits, which led them to cause a tragic death, for example, by not knowing, despite inquiries, that there were innocent civilians where a bomb was being dropped, then it is for someone who needs to come to terms with the implications for their humanity of deliberately killing an innocent person, like a soldier who has surrendered. It is the ability to help service personnel recognize and make these kinds of distinctions that is particularly important to the task of military ethics, to reduce the risk of unnecessary moral injury, or to provide tools to assist in recovery from it.

While the thinking behind why we focus on intention may have some complexity, the advantage in practice for military ethics is that this first step of natural law theory can be distilled to a simple and intuitive question, 'is my intended outcome good?' If someone is wondering what good means, they can ask the question in a slightly different way, 'for the people I am trying to help will my intended actions make their life better?'

What Are Reasonable Means

Having asked, 'is my intended outcome good?', we need to ask, 'is my plan to pursue that end reasonable?'

There is no simple formula to define what reasonableness is. Rather, as John Finnis puts it, reasonableness is what is required to pursue a good end in the face 'of human wants and passions and the conditions of human life' (Finnis 1991, 101). For example, we know that one of the conditions of human life is that we live in a world of limited resources, so if we have to choose between two ways of doing something, all other factors being equal, it is reasonable to choose the more efficient way.

As Finnis observes, throughout history philosophers and ethicists have set out various approaches to creating general principles as to what is reasonable (Ibid., 102). Many illuminate aspects of what it is to be reasonable in pursuit of your goals even if some like Immanuel Kant would see those principles as sufficient in themselves and needing no theory of ends.

For example, Kant's (1785) famous maxim 'act in such a way that you treat humanity, whether in your own person or in the person of any other, never merely as a means to an end but always at the same time as an end' might stand on its own, but it is also a very sound principle as to how to treat people reasonably in pursuit of good ends (Finnis 1991, 122). In a military context, that principle provides a sound basis for why in the pursuit of the good end of international justice it is not reasonable to use as means, actions like taking hostages, threatening the mass killing of innocent civilians to force a surrender, or torturing people.

In complex ethical situations such as those thrown up by war, the most helpful principles of reasonableness are often at the next level down from these sweeping general principles. They are the sorts of principles that provide more specific guidance about how to deal with particular challenges. It is a feature of natural law theory that it moves from general principles of reasonableness to norms that guide decisions.

In essence, just war theory is largely just that. A distillation of what reasonableness means when it faces that perennial condition of human life, remedying serious international injustices.

If we consider some of those principles for *jus ad bellum*, we can see how they are grounded as a test of reasonableness. Take the principle of last resort for example. That starts with the fact that our end is to remedy an international injustice. We have already seen that a principle of reasonableness in a world of limited resources is to choose the more efficient means to an end, all other factors being equal. Similarly, if we can remedy the injustice, all other considerations being equal, in ways that cause less harm, we should do so. Given that war in all probability is going to cause more harm than any other way of remedying the wrong, it should be our last resort.

If we turn to the principles of justice in war (*jus in bellow*), especially the core ideas of acting with discrimination and proportionality, the feature of natural law theory of turning a broad set of principles that are useful for analysing a dilemma into simple but practical norms to guide decision-making is very evident.

Consider the just war requirement of proportionality. It is derived from the larger general idea about reasonableness when it comes to causing harm known as the principle of double effect. It is an analytic insight that Thomas Aquinas is generally credited for introducing into Western thought in his discussion of self-defence when he observed:

> Nothing hinders one act from having two effects, only one of which is intended, while the other is beside the intention. . . . Accordingly, the act of self-defence may have two effects: one, the saving of one's life; the other, the slaying of the aggressor.[7]

The logic of this idea is grounded in the centrality of intention in natural law theory. In these situations what matters first of all is that a good end is intended. In this case, what is intended is the saving of one's own life, not the killing of an aggressor. Part of how we know that saving one's own life, not killing the aggressor, is the real intention is to ask, 'would our plan be successful if our life was saved and by some good fortune the aggressor also survived?' If the answer is yes, then it is clear that person's death wasn't part of the outcome we intended. This answer also explains why we aid wounded soldiers on the opposing side who are no longer able to fight and take as prisoners of war those who have surrendered or are unable to fight any longer.

However, when in self-defence someone does die or is injured, the ethical question is not closed. This death that is a side effect of our choice is still of great ethical and existential significance. Someone's death has after all followed from our actions. We are involved in the causal chain of serious harm. The question becomes whether, although unintended, it is even reasonable or fair to accept that side effect.

It is here that the test of proportionality helps to determine what is reasonable or fair in the circumstances. There are two elements to that test of proportionality.

The first, in this self-defence case, or any similar case of the use of force in war, the question is whether the force used was the minimum necessary to achieve the intended effect. To be clear, the intended effect is not the death of the person but stopping them from killing you. If that was possible with non-lethal force and you killed them, then it would be a disproportionate act.

The second element of the test is whether there was broad symmetry between the good end that was intended (e.g. saving someone's life) and the harmful side effect (e.g. the death of the assailant). In this case with two lives, albeit two incomparable lives, that symmetry is apparent. If to save that one life it was necessary to kill a significant number of people in addition to the assailant, there would be no sense of balance or proportionality. This isn't a utilitarian analysis because there is no exact number of lives that determines where the line is between proportionate and disproportionate. If we respect the true uniqueness of human lives then they can't be compared to commodities that can be exchanged for one another. Equally, no one life can be measured to be worth many other lives. That is why the respect for human uniqueness and the incomparable nature of different types of harm requires evaluation in terms of a broad symmetry.

This is where natural law theory respects that ethical judgement is a form of human judgement and is bounded by human limits, which means we simply can't compare the incomparable. This is an important point for military ethics because it introduces space for the fact that different soldiers and their commanders may have varying judgements about what constitutes proportionality in a particular situation and that in assessing those judgements we need to bring a significant measure of humility to them and to recognize that there can be within limits a range of ethically acceptable answers.

If we turn to that important task of military ethics of practical decision-making by military personnel in action, we can distil all this logic down into a much simpler requirement, 'in taking action against the enemy, be proportionate'. In practice, it means that a soldier on the front line doesn't need to do a lot of complex ethical analysis. Rather, in most combat situations they just need to make sure that they are targeting the enemy (i.e. being discriminate) and using proportionate force to stop them or overcome them.

Of course, there will be occasions when a more difficult choice or set of choices present themselves. For example, is targeting a railway used to supply the front line discriminate? Even here, within the layered way natural law thinking operates, even a junior officer with modest ethical training will know to ask the next level of question to reveal whether the action is discriminate, in this case 'is the railway line I am targeting an integral part of the way in which the injustice of the war is being committed or is it being used much as it would have been in peacetime?' So, if the train line is the main supply line to the front line then clearly it is a legitimate target but if it is part of a commuter network that trains carrying armaments are incidentally crossing, it is not.

This is the power of natural law theory's analysis. It can be used in both shorthand form to deal with familiar changes and longer hand forms when more complex analysis is needed. In military settings, this is essential because you can't train all frontline military personnel to be ethicists, but you can train them all to use the shorthand guidance that is both related to and on a par with rules of engagement and laws of armed conflict. Equally, you can train officers who will have specific tasks of ethical significance like targeting with the more elaborate thinking tools they need but even this knowledge is only going to be on par with the sort of legal knowledge that one would regularly receive anyway for those roles.

Before going further, it is important to note the existential not just the ethical value of this sort of analysis. We identified being able to help with the existential dimensions of being involved in conflict as one of the tasks of good military ethics, which can reduce the risk of unnecessary moral injury. The type of analysis of reasonableness we have started to unpack here gives a sense of how natural law theory can provide those involved with important perspectives to explore the moral character of their experience. The fact that killing people isn't the objective of conflict means they don't need to have the character of a killer. That some deaths of innocent people can be tragically accepted as the side effects of good ends not worn as wrongful acts. Beyond those tragic deaths that can be reasonably foreseen in war, there will also always be tragic deaths that can't be foreseen. These are deaths that can be viewed as part of accepting that there are limits to what humans can know and decide and accepting that we can't judge ourselves by some more than human standards. The greater a person's ability to accurately characterize the ethical and existential qualities of their experience, the less risk there is of unnecessary moral injury. Those preventable injuries arise when someone believes they transgressed an important moral norm when in fact carefully considered, they haven't.

In this discussion, we have focused on a key application of the general principles of reasonableness to the situations requiring norms of proportionality and discrimination. It is important to recognize that just war theory as well as natural law theory provides powerful ethical thinking tools for other ethically challenging domains of military activity.

To give a sense of the breadth of tools available, it is worth highlighting an example of the analytic approaches available for the gathering and use of military intelligence, which can also be an ethically complex area. For example, can you use an intelligence product, which might save many lives, that includes, amongst other sources, information obtained by a partner country using unethical methods? The question of whether it is reasonable to use that intelligence product involves asking whether by using it you are participating in the wrongdoing? Natural law theory has developed thinking tools to tackle the issues of possible participation in wrongdoing (Finnis *et al.* 1987).

Again, the answer starts with some important distinctions about intention. If you share the unethical intention of the person doing the wrong and are seeking to benefit from it, then you are engaging in formal cooperation with the wrongdoing, and it is unreasonable. For example, if you knew prospectively that the partner country was going to torture people and you were planning to use the information they obtained that would be *formal cooperation*.

However, if the information has already been obtained you don't need to share their intention about how they obtained it. Indeed, you can obtain it and coherently say that you think it was wrong the way it was obtained. In these sorts of cases, you are only *materially cooperating*. Whether you can reasonably cooperate materially depends on the side effects of that cooperation. In this case, if using the product contributed to encouraging the partner country to continue their unethical practices or supported a culture in your own intelligence operation of unquestioningly using unethical material sourced from others then it might amount to unreasonable material cooperation. However, if incorporation of that sort of information was relatively uncommon and using it was unlikely to have any impact on the partner country's practices then given the good ends (saving lives) to which it will be put means the material cooperation need not be unreasonable.

Hopefully, this example illustrates not only the sort of sharp analytic tools available but also that with such tools we don't need to resort to the use of utilitarian theories when faced with a tough choice like this of saving many lives versus torturing one person.

In this section, we have dealt with just two of the analytic tools natural theory makes available to help think through what is reasonable in various military or other settings. There are a range of others covering issues from privacy and promise-keeping to truth-telling and conscientious objection. It is that range of thinking tools that provide much of the power and practicality of natural law theory.

Principles of Reasonableness and Duties

Having considered how principles of reasonableness can operate in military settings, it is worth observing that acting consistently with such principles can be expressed as duties. The consistent application of the principle 'not to use people as a means to an end' when interrogating them can be expressed as a duty not to torture people. Acting discriminately to ensure you are always focused on the good end of remedying an injustice can be expressed as a duty not to intentionally harm innocent people.

To make ethics usable, being able to express at times complex principles in terms of simple duties can be very helpful. The construct of duties is also a powerful tool for constructing ethical cultures, which themselves help to sustain ethical conduct.

Sometimes people seek to contrast an ethics of duty (deontological ethics) with an ethics of ends (teleological ethics) like natural law theory. It is a false dichotomy because natural law theory is both an ethical theory of ends and of duties, which arises from it being concerned with questions of both the reasonableness of our intended ends and means.

Principles of Reasonableness and Virtues

The consistent application of principles of reasonableness is not just duty forming they are also character forming. In the Australia Defence Force (ADF) Military Ethics Doctrine, this relationship between principles and character is made particularly clear (ADF 2021, 29–30). The Doctrine identifies that each of the ADF's values, which include qualities like Courage and Integrity, are in fact virtues or qualities of character. Significantly, that doctrine then identifies that in the ADF's values system there is a principle associated with each of these virtues and that it is acting consistently with those principles (highlighted in the Doctrine with italics) that leads to a person having those qualities of character. So, for example, with 'Courage' it is '[t]he strength of character *to say and do the right thing, always, especially in the face of adversity*' (Ibid.) and with 'Integrity', it is '[t]he consistency of character *to align my thoughts, words and actions to do what is right*' (Ibid.).

Contemporary natural law theory sees this sort of development of virtue as an important feature of how it operates. Virtue for the contemporary natural law theorist is not just a by-product of consistent decision-making, it also guides future decision-making. Germain Grisez explicitly articulates how that relationship works:

> Once the choice is made, a certain aspect of one's self is involved in the good one has chosen, which is not what is involved in the alternative. One is as one has chosen to be. If the very same alternatives were to present themselves again – everything one judges good or bad being the same – one could have no reason for choosing otherwise. And so no new choice would be necessary. That is why previous choices provide fixed points of reference to resolve further situations without new choices.
>
> *(Grisez 1983, 51–52)*

To explain the depth of contemporary natural law theory as a virtue theory, it is important to highlight that it is the whole self, both the rational and emotional self, that is involved in the decision (Grisez *et al.* 1987, 122–125). While to this point in the chapter, we have focused on the rational part of natural law theory, the theory recognizes that emotions are integral to decision-making because it is the feelings associated with particular good ends that make them attractive or unappealing. For example, to explain the choice to pursue a friendship, there is both a question of logic but also of the anticipated or existing feelings associated with that friendship.

Part of why it is helpful specifically to identify natural law theory as a sophisticated form of virtue ethics is because of the military concern with the overall formation of character. That means that forming ethical military personnel isn't a task over to one side but can, and should be, an integral part of the project of developing and maintaining, in the fullest sense of the word, a *good* soldier, sailor or airperson.

Central to that task of creating character is culture. Culture at its simplest tells people 'how we do operate around here'. In conforming to that culture, a person's character is shaped and along with it their propensity to make good or bad decisions. We know about the importance of this connection often most clearly when unethical acts occur because almost inevitably the inquiry into them becomes at least in part an investigation into the culture that led up to those events.

The Legitimacy of Authority

So far, we have focused on individuals making their own decisions but one of the key features of ethical decision-making in military contexts is that people are commonly executing other people's decisions. It begins with instructions from political authorities to engage in military action and continues through the orders issued right down the chain of command. In executing those orders, there is commonly substantial scope for interpretation about how an officer or soldier seeks to implement the commander's intent but nevertheless the basic requirement that they 'follow orders'.

The obligations and boundaries to following orders require us to answer, 'why is it reasonable to obey authority in the first place and when and why would it not be reasonable to obey authority?' Here we are putting the question in a form that makes it clear we are dealing with an act of authority not just an agreement. The question is why should we follow a direction from someone else, be it in the form of a direct command, a law or regulation, or a court judgement irrespective of whether we agree with it or not? Natural law theory has devoted a lot of analytic attention to this question with the most substantive contemporary treatment of it being John Finnis' seminal work *Natural Law and Natural Rights*, on which this analysis is based (Finnis 1991, 231–259, 351–367).

The reasonableness of obeying authority starts with the idea that it is a principle of reasonableness to seek to pursue the common good. The common good are those life-enhancing outcomes that we can only have by cooperating with others. At the simpler end, it is the ability to work with others to create shelter or a house we couldn't build alone, and at the more complex end, it is the delivery of modern health care, which from the training of medical staff to the development of drugs, requires extraordinary levels of cooperation.

The key quality to realising any of these aspects of the common good is cooperation. Without it, the many benefits of working together couldn't be realized. Those benefits from housing to health care are very substantial so we have a strong reason to cooperate.

The challenge comes when we disagree, for example, about how to build the house or how to run the health care system. In these situations, we can face a choice, for example, between not cooperating and not having a house built or cooperating and having the house built despite the fact we think there is a better way to do it. Given the substantial good that comes from having a house, we have good reason to cooperate with a plan we don't agree with.

In any larger group of people, we will never get a consensus about what the common good should be (e.g. what mix of housing, health care, education, economic growth we would seek) or how we should pursue it. Yet, the only way the common good will be realized, is if there is cooperation. We have to establish a way in which these cooperation problems are resolved. That can occur in as many forms as there are types of government and organizations. These governments and organizations that solve the cooperation problem do so because they exercise authority. In other words, people recognize the reasonableness of following their directions to realize the benefits to society that come from being organized even when they don't agree with the particular directions.

In military settings, the general rationality of following orders is particularly clear. The chance of a successful military outcome is very small if individual soldiers or units make up their own minds about whether or not they follow the plan. The potential consequences of not following orders also make it very clear what a serious matter it would be not to do so. That underpins the strong obligation on military personnel to follow orders.

Nevertheless, there are situations where reason requires military personnel to disobey an order. Those are situations where laws or orders are directed against the very reason why we have orders or laws in the first place, which is to advance the common good. An order or law aimed at attacking a civilian population and destroying their society and their common good along with it or arbitrarily denying their participation in the common good whether by means like deportation, enslavement, or imprisonment might have the form of a law or order but not the substance.

The reason it would not just be reasonable but rationally required not to obey these sorts of orders is that there is not only no common good to be advanced but it is being destroyed so the alternative, which is a breakdown of order among those committing these acts may in fact be the only option that enables the common good to be protected.

It is always important to ask whether the side effect of not following an order or obeying a law can reasonably be accepted. A question that always needs to be asked is whether in not obeying a law or order someone is weakening the general readiness to obey laws. However, in the situations we are describing here, there is little risk of this sort of disobedience weakening the general effectiveness of military orders. Rather, it is likely to weaken the likelihood that those with military authority will seek to issue such orders in the future because it will strengthen the possibility that they will be disobeyed.

It is worth observing, given the tasks we set out for military ethics, that the rationale we have set out provides the ethical logic for why military personnel can't avail themselves of the defence of superior orders or the prescription of law for acts of genocide or crimes against humanity as per Article 33 of the Rome Statute of the International Criminal Court.

Similarly, at a more tactical level an order to kill a non-combatant, to murder them, is an order that a soldier is obligated not to obey. First, just as in cases of genocide and crimes against humanity, the order is not authoritative because it contradicts the very reason for having an order which is to enable a person to participate in the common good. Second, given the law is not binding it is wrong (practically unreasonable) to intentionally kill innocent people.

The same logic will apply to any other orders to commit crimes against the person from torture to the harsh and degrading treatment of prisoners of war – the type of war crimes prohibited in the Geneva Conventions and in Article 8 of the *Rome Statute of the International Criminal Court.*

If we step back to consider both the strong general obligation on military personnel to obey orders along with the specific situations in which that obligation doesn't hold, we can make some further observations about the operation of the authority in military settings.

The first are, that unless an armed conflict clearly constitutes a campaign whose objective or essential strategy is genocide, crimes against humanity or the commission of war crimes it is reasonable for members of the military to follow orders to participate in an armed conflict. If it subsequently turns out that it was not a just war, then the responsibility rests with those who made the decision to go to war not with the military personnel who followed the lawful directions of those decision makers.

The second is that in situations where a member of the military has reason to be concerned that their orders may be unjust, for example they are concerned the targeting they have been given might not be discriminate or proportionate, the reasonable response is to raise the concern their chain of command. If having raised it, the order is still to proceed and they have no wider reason to think their chain of command is given to acting in breach of the laws of war, then it is reasonable to follow those orders. If, however, they reach a point where on a reasonable assessment of the evidence think they are being ordered to take actions they are obligated not to do, then, if circumstances permit, they need to signal that they won't comply before any final refusal of the orders. In doing so they need to know that given the seriousness of maintaining order, very significant consequences for disobeying them may follow.

Third, there is a wider importance in democratic societies to ensuring that the military has a deep ethic of following lawful directions because that is a major force that helps to ensure that the military supports the rule of law and with it civilian control of the military.

Natural Law and Tasks of Military Ethics

At the start of this chapter, we identified a series of tasks that military ethics needed to perform and four features any theory would need if it was to fulfil those tasks. We can now assess natural law theory against those features.

The first of those features was a theory that had the tools to analyse complex problems but was also a virtue theory that could help with the shaping of culture and guiding of daily decisions. By explaining the integral connection between the operation of principles and the creation of virtues in natural law theory, we have seen how and why it can be such a theory.

Second, the theory needed to provide simple effective guidance in tough high-pressure ethical situations such as those faced by soldiers in combat but also be capable of analysing complex ethical dilemmas. We have seen that in applying general principles of reasonableness to a range of challenges that are integral to the human condition, natural law theory has produced a series of thinking tools like just war theory, which provide both rich, comprehensive, and consistent ethical analysis but also simple, usable, principles like acting discriminately and proportionately. We have noted that many of these thinking tools are particularly helpful in dealing with situations faced by people in the military. We observed that this matters not just to ensure military personnel act ethically but also to reduce the risk of unnecessary moral injury.

Third, we saw how natural law theory provides a sophisticated account of authority that can both explain the general ethical obligations to obey laws and orders and be clear as to when a law or an order should not be followed.

Finally, we saw the need for consistency between the theory's ethical solutions and international law, especially the laws of armed conflict, and the broad premises of domestic laws as well. We saw that one of the strengths of natural law theory is precisely that it is underpinned by the tradition out of which the modern laws of armed combat emerged and international law more broadly.

We also considered the ethical logic that requires soldiers to obey commands generally but to refuse to obey orders when obeying them would involve them in committing criminal acts where the defence of superior orders isn't available to them under international law. That is very helpful because it means soldiers aren't going to be caught in that terrible dilemma between doing what is right and doing what is legal.

That broad consistency between natural law theory and domestic law, at least in the common law world, also holds true because the common law is a form of law whose roots are deep in the natural law theory. That relationship is explicitly described in the important founding text of the English Common Law, Sir William

Blackstone's *Commentaries on the Laws of England* in which he observes the law of murder:

> To instance in the case of murder: this is expressly forbidden by the divine and the natural law; and from these prohibitions arises the true unlawfulness of this crime. Those human laws that annex a punishment to it, do not at all increase its moral guilt, or superadd any fresh obligation.[8]

As Blackstone also makes clear, we wouldn't of course expect to see natural law theory underpinning all laws because there are a wide range of issues from regulating international trade to licensing architects that the common law and statutory laws regulate which don't look to fundamental ethical principles to ground them (Ibid.).

Making It Work in Practice

We could summarize the argument in this chapter by observing that in relation to military ethics, but in reality in relation to all applied ethics, an ethical theory must pass both philosophical and practical tests.

A theory must be philosophically robust and sufficiently sophisticated to provide an intellectually satisfying answer to the breadth and complexity of ethical issues we face. In this chapter, we have highlighted how natural law theory, which is built around inquiry into the practical reasonableness of our intended ends, provides a single theory that can coherently integrate ethical concepts of ends, principles, duties, virtues, and consequences. As we have seen, these concepts are all important for ethical analysis. Too often we are asked to choose between theories that are built principally around one or two of these concepts or a logically problematic hybrid that seeks to bolt multiple theories together to create a more complete account.

An ethical theory also has to be usable for day-to-day decision-making by everyone. Thanks to its long history and deep engagement with military matters, it is a very conceptual rich and well-developed theory. However, it is also eminently practical. There are just two basic questions people need to ask in most situations:

- Is the end I am intending a good one?
- Are the means I am using reasonable?

In a military context, if someone is wondering, 'what is reasonable?' they can ask how does this fit with the military's values. If they are a well-constructed set of values underpinned by principles of reasonableness, like the ADF's, then asking whether I am acting with 'respect', 'courage' and 'integrity' will provide a very good orientation to what being reasonable looks like.

For military personnel involved in the planning or execution of military operations, they will need to acquire some specific ideas to help them think through what is reasonable. In this case, it will be ideas about the two key concepts of discrimination and proportionality. In supporting people to have facility with these ideas, we are doing no more than what is needed to ensure they understand the laws of armed conflict, which is a plausible ask. Indeed, in this case because of the positive relationship between ethics and law, providing training on both these issues will be mutually reinforcing.

In other areas of military life, such as intelligence gathering, there will also be some specific thinking tools that people need to acquire. What this recognizes is that as in any area of professional life, there are specific ethical skills that individual occupations need, to ensure they apply the general ethical principles to the specifics of their work. Given that the military involves a range of occupations brought together to enable military effects to be created, it is to be expected that there would be a range of different thinking tools needed.

Importantly, for natural law theory, where both practical reasoning and the duties it gives rise to are also virtue forming, creating an ethical military is not a task on its own. Rather, it is an integral part of that core larger task of any military organization, which is to create the culture, character, and conduct that support successful military operations. That is especially true when being ethical confers not just a moral advantage but a strategic one.

Notes

1 Jean Barbeyrac's 1735 translation into English published in 1738 is the most important and influential edition. It has been reprinted as *The Rights of War and Peace*, Books I – III, edited and with an introduction by Richard Tuck (Indianapolis, IN: Liberty Fund, 2005).
2 Key foundational texts include Grisez *et al.* (1987) and Finnis (1980). For a short introduction see Nigel Biggar and Rufus Black, *The Revival of Natural Law* (Routledge, 2000). 'Introduction: The New Natural Law Theory.'
3 Aristotle, *Nicomachean Ethics*.
4 For an account of what constitutes these ends and a comparison related to human well-being from other schools of philosophy as well as anthropology and development economics see Alkire (2000).
5 This notion of cultural embodiment is explored more fully in Alkire and Black (1997, 263–279).
6 In these sorts of cases, it is also possible that someone is acting in pursuit of some other end worthwhile, e.g. killing an innocent person to terrorize others, so they will stop posing a threat. When that is occurring the unreasonableness lies in the means intended and we will deal with that below.
7 Thomas Aquinas, *Summa Theologica* (1485, II-II, Qu.64, Art.7)
8 William Blackstone, *Commentaries on the Laws of England, Volume 1 A Facsimile of the First Edition of 1765–1769* (The University of Chicago Press, 1979), Introduction, Section 2: On the Nature of Laws in General.

References

Alkire, Sabina. 2000. 'The Basic Dimensions of Human Flourishing: A Comparison of Accounts' in Nigel Biggar and Rufus Black (eds) *The Revival of Natural Law*. Routledge.

Alkire, Sabina, and Rufus Black. 1997. 'A Practical Reasoning Theory of Development Ethics: Furthering the Capabilities Approach.' *Journal of International Development* 9, 263–279.

Aquinas, Thomas. 1485. *Summa Theologica*.

Aristotle. *Nicomachean Ethics*. Translated by Terence Irwin 1985, Hackett Publishing Company, Indianpolis/Cambridge.

Australian Defence Force. 2021. ADF-P-0 Military Ethics, Edition 1. Directorate of Information, Australian Defence Force. https://theforge.defence.gov.au/sites/default/files/2021-12/ADF-P-0%20Military%20Ethics%20Ed%201_0.pdf

Biggar, Nigel, and Rufus Black. 2000. *The Revival of Natural Law*. Routledge.

Blackstone, William. 1979. *Commentaries on the Laws of England, Volume 1: A Facsimile of the First Edition of 1765–1769*. The University of Chicago Press.

Finnis, John. 1980. *Natural Law and Natural Rights*. Clarendon.

Finnis, John. 1991. *Natural Law*. New York University Press, Reference Collection.

Finnis, John, Joseph Boyle, and Germain Gabriel Grisez. 1987. *Nuclear Deterrence, Morality, and Realism*. Oxford University Press.

Grisez, Germain Gabriel. 1983. *Christian Moral Principles*. Franciscan Press.

Grisez, Germain Gabriel, Joseph Boyle, and John Finnis. 1987. 'Practical Principles, Moral Truth, and Ultimate Ends.' *The American Journal of Jurisprudence* 32(1), 99–151.

Grotius, Hugo. 1625. *De Jure Belli Ac Pacis Libri Tres*. (Note: Jean Barbeyrac's 1735 translation into English and published in 1738 is the most important and influential edition.)

Grotius, Hugo, Jean Barbeyrac, and Richard Tuck. 2005. *The Rights of War and Peace Bk 3*. Liberty Fund.

Kant, Immanuel. 1785. *Groundwork for the Metaphysics of Morals*. Oxford University Press.

Litz, Brett T., Nathan Stein, Eileen Delaney, Leslie Lebowitz, William P. Nash, Caroline Silva, and Shira Maguen. 2009. 'Moral Injury and Moral Repair in War Veterans: A Preliminary Model and Intervention Strategy.' *Clinical Psychology Review* 29(8), 695–706.

Phelps, Andrea J., A. B. Adler, S. A. H. Belanger, C. Bennett, H. Cramm, L. Dell, D. Fikretoglu, et al. 2022. 'Addressing Moral Injury in the Military.' *BMJ Military Health*.

PART II
Analysis and Critique

6

BAKER RESPONSE

Deane-Peter Baker

I am very grateful to my co-authors for their efforts in articulating what, in their view, is the best model for military ethical decision-making. Each has made a strong case, and in so doing has advanced the cause of military ethics a step further. But it is the intention of this book to go beyond simply offering options, and so I turn now to offering a respectful critique of the models that they have put forward.

Let me begin with the account Roger Herbert has given us of the 'Moral Deliberation Roadmap' that he and his colleagues developed to impart to US Naval Academy Midshipmen. It will not surprise the reader that I find myself in hearty agreement with almost everything in Herbert's chapter. As mentioned in my opening chapter, I first began teaching the 'Ethical Triangulation' ethical decision-making model when I was myself teaching *NE203: Ethics and Moral Reasoning for Naval Leaders*, in my former position as a member of the Department of Leadership, Ethics and Law at the US Naval Academy. This was prior to the review of the curriculum that Herbert discusses in his chapter, but introducing Ethical Triangulation into the classroom was my way of addressing the 'moral theory smorgasbord' problem with the curriculum back then. It is unsurprising, then, that there is so much resonance between the latter steps of the USNA Moral Deliberation Roadmap and the Ethical Triangulation model I have presented.

I am also delighted that the Roadmap highlights the importance of moral perception, as well as the impact that such factors as time, environment, and personal biases can have on ethical decision-making. Like most military ethicists I started my academic life as a philosopher, and consequently, that was for some time the primary lens through which I viewed the ethical challenges of military operations. However, the more time I have been able to spend in conversation with experienced military practitioners, the more I have come to realize the importance of what is sometimes referred to as 'behavioural ethics', and this has come to take up

DOI: 10.4324/9781003312925-8

a significant portion of my military ethics teaching. When, in 2020, I was asked by Australia's Special Operations Command (SOCOMD) to assist them in developing a coherent plan for training and educating their personnel in ethics, I proposed a model of 'ethical fitness' that incorporated the following elements:

1 Ethical knowledge,
2 Ethical perception,
3 Ethical reasoning,
4 Ethical motivation, and
5 Ethical 'grit'.[1]

Of these, only the first element is specifically associated with ethical or moral theories, but without the other elements being in place even the most well-educated soldier[2] cannot be relied upon to behave ethically. The NE203 curriculum, in my view, does an outstanding job of addressing these broader factors.

But I do have a concern about the Moral Deliberation Roadmap – or perhaps it is simply an as-yet unanswered question. Are seven 'steps' too many? I do not, in raising this question, intend in any way to suggest that we should not be preparing soldiers to develop ethical perception or to understand and mitigate the impact of bias and environmental factors on our decision-making. We absolutely should, and I enthusiastically do![3] But as I averred in my previous chapter (when responding to Benjamin Ordiway's critique of ethical triangulation), I tend to address the 'system one' and 'system two' challenges of ethical decision-making separately in my teaching. This allows me to distil the conscious deliberative process of decision-making into three concise steps. As Rufus Black rightly notes in his opening chapter, one of the things we want is to 'enable military personnel engaged in armed conflict to make very rapid, reliable, and consistent ethical decisions under conditions of extreme pressure'. Given this, it would seem that – on the face of things at least – the fewer steps we ask soldiers to go through in those moments, the better.

Of course, it would be simple enough to teach military personnel to skip directly to step four of the Roadmap when circumstances dictate, and I imagine that is what happens in the delivery of the revised NE203 curriculum. The difference between the Roadmap and Ethical Triangulation at that point is small – just one additional step. In my view, the distinction in the Roadmap between 'upper-tier' factors (constraints and consequences) and 'lower-tier' factors (special obligations and character) is both conceptually and pedagogically useful. The inclusion of a step requiring consideration of special obligations unquestionably gives the Roadmap greater resolution than Ethical Triangulation. In Ethical Triangulation, special considerations are lumped in with the other 'constraints' the decision-maker is asked to account for. The key question, it seems to me, is whether or not *in practice* the advantage of clarity gained by having this additional step outweighs the (presumed) advantage that Ethical Triangulation has in terms of speed and simplicity? I simply have no idea. The good news is that this is a question that is at least in principle empirically testable. My hope is that this is exactly the sort of empirical work that we in the field of military ethics will start to do. One thing I *do* know is that courses like NE203 are a luxury. Outside of service academies like USNA and ADFA, the likelihood of military personnel having the opportunity to spend up to 16 weeks being educated in ethics (as in the case of NE203) is vanishingly small. The vast majority of soldiers – those who are not commissioned officers – are unlikely to receive more than a few hours of education on this topic over their entire careers. This, it seems to me, is another point in favour of having a model that is as simple as possible while remaining practically useful.

I turn now to consider the ethical decision-making models put forward in this volume by Rufus Black and Iain King. Black's natural law-based approach and King's quasi-utilitarian Help Principle are clearly very different to one another, and so it may at first seem odd that I am responding to them together. However, as I will show, they share some key features in common, and it is these that I wish to focus on in my response.

The first important thing these two approaches have in common is that they both have very specific metaphysical and ontological foundations. King takes considerable care to ground the Help Principle in a broadly Darwinian account. Black's natural law-derived model is, obviously, grounded in natural law ontology. Neither scholar, I think, would wish to separate their ethical decision-making model from its underpinning philosophical assumptions. I think it is also fair to say that these foundations are irreconcilable – while there have been some valiant attempts to draw together Darwinian evolution and natural law,[4] most philosophers would agree that they are deeply incompatible. Natural law thinker Bernard Mauser helpfully outlines some of the key differences as follows:

It takes very little time when researching ethical theory to find that natural law has fallen into disrepute in the American academy. Perhaps part of the reason is the influence of Darwinism and the ideas that follow from it. It may be that Darwinism has motivated the development of theories which deny several things necessary for natural law theory. One such denial is the existence of a fixed human nature. Darwin's theory explains that life begins in something like primordial soup, and eventually evolves from simple to complex life. There are no "fixed" unchanging essences that stabilize a species, as one species gives rise to other species. Dismissive of any essence or nature, those following this view state that there is no basis for establishing an ethical theory on a fictional thing like human nature. Another denial inherent in Darwinism is that there is no teleology or final causality in nature. Human nature, teleology and final causality, which were embraced in an Aristotelian framework, were replaced by mechanism and the search for purely material causes in the realm of science. Because of these things, there is only an "appearance of design" in organisms which is simply illusory.

(Mauser 2011, 2)

Which of these competing metaphysical accounts should we choose? Most of us will have a leaning one way or another. Though not religious in all its forms, natural law theory has its deepest roots in Christian theology – as Mark Murphy notes in the *Stanford Encyclopaedia of Philosophy*, 'If any moral theory is a theory of natural law, it is Aquinas's'.[5] Consequently, those whose personal beliefs are religious in nature are more likely to find the natural law approach appealing. If you are in this camp, you will also likely find King's view that 'our ethical attitudes are mostly the product of an arbitrary evolutionary process' deeply problematic. And you will likely be uncomfortable with some positions associated with broadly utilitarian approaches to ethics, such as the claim that, in some circumstances, infanticide may be morally justified.[6] On the other hand, you may be someone who finds in Darwinian theory all the metaphysical explanatory power you need. In that case, you will likely find the association of natural law thinking with religion to be unappealing, you will likely be uncomfortable with its association with colonialism,[7] and you may be troubled by the way some scholars have linked natural law's idea of human nature to arguments against abortion, same-sex marriage, and the like.[8]

As I have tried to argue in my opening chapter, it seems to me that asking 'which of these competing metaphysical accounts should *we* choose?' is precisely the wrong question when considering which ethical decision-making model is appropriate for the military arm of the state. The idea of liberal neutrality precludes the state from committing to any particular metaphysical or ontological account. This is why Ethical Triangulation (and the Moral Deliberation Roadmap) does not 'pick a side'. Instead, the armed servants of the state are asked to consider ethical quandaries by – as best as is practically possible and operationally effective – weighing the primary intuitions underpinning the most influential ethical approaches expressed in our society.

The concern over 'picking sides' is exacerbated in the case of the models put forward by Black and King in that they are both *outliers*, philosophically speaking. While natural law thinking has unquestionably been hugely influential in the history of Western thought, that influence has (as the natural law thinker Bernard Mauser noted earlier) waned significantly in contemporary academic and political discourse, with conceptions of 'human rights' largely displacing the idea of 'natural rights'. As James Rachels wrote in the best-selling ethics textbook of the past 30 years, *The Elements of Moral Philosophy*, 'Natural law theory plays virtually no role in contemporary secular moral discourse'. (Rachels 2007, 60)

A possible rejoinder here is to point out, as Black docs in his opening chapter, the connection between natural law theory and the principles of the just war tradition. Perhaps it might be argued that, though natural law theory is somewhat peripheral to contemporary ethical discourse, that is less relevant because we are talking here about ethics for the military, and the core principles guiding the use of military force are those of the just war tradition. I am not sure I agree with this line of thinking. Military professionals already have the principles of the just war tradition to guide them, they do not need a general decision-making model to do the same work – it is sufficient, in my view, that the two are compatible (as I have shown Ethical Triangulation to be). Still, it is certainly true that figures like Aquinas and Grotius are foundational to the just war tradition *and* central to natural law thinking. However, care must be taken not to misconstrue the nature of the just war tradition. As Valerie Morkevicius writes in her excellent account, it is

> a tradition of practical reasoning, emphasizing the analysis of real cases more than the refinement of principles. . . . As a tradition – a sort of common language – it "gathers together the learning of previous generations", to help us ask the right questions [to] guide our decision-making in the present.
>
> *(Morkevicius 2018, 20)*

Natural law thinkers have unquestionably played a vital role in shaping that tradition, but so have others. For example, the most influential just war thinker of the last 100 years is unquestionably Michael Walzer – in just war tradition terms, the Grotius of our times – who is no natural law thinker. As Morkevicius explains, contemporary just war scholarship boasts of three quite diverse perspectives. The first are the 'neoclassical' thinkers, those who

> draw on the history of the Western just war tradition (and sometimes non-Western traditions) to think critically about how the classical just war principles can be interpreted and reinterpreted to analyse the moral problems of contemporary war. . . . [I]n this view our "conversations with past generations" can help us recognize our own parochialism, reminding us that our own situatedness can limit our imaginations.

The second group are Walzerian,

> those who follow in Michael Walzer's footsteps, referring broadly to the historical tradition, but fundamentally deriving their ethical conclusions from their moral intuitions. . . . Walzer's *Just and Unjust Wars* revitalized just war thinking in the West, melding the traditional Christian language of the just war criteria with secular legal and philosophical arguments. Like neoclassical just war thinkers, the Walzerian approach is collectivist, emphasising groups rather than individuals. Indeed, like structural realism, the Walzerian approach is decidedly statist.

Finally, there are the newcomers, the *revisionists* or reductionists [who] approach just war thinking from a radically different perspective. Although they use the language of just war, revisionists have radically reimagined the logics of just war thinking by reasoning from the individual up (Morkevicius 2018, 21–22).

Of these groups, it is the neoclassicist approach that is most closely aligned to natural law thinking, given its historical focus. But they are also a small and dwindling group in the academy, with Walzerians arguably holding the centre and revisionists growing in number and influence. Thus, even if we narrow our view from contemporary ethics to contemporary just war thought we find natural law thinking somewhat on the periphery.

King's quasi-utilitarian approach seems, at first glance, to be more central to contemporary ethics discourse, being based as it is in evolutionary theory and drawing on utilitarianism. But the ambitious and interesting 'Help Principle' that King derives from these sources is of his own making and is hence also an outlier. Even if one thinks, individually, that natural law or the Help Principle offers the most philosophically coherent and practically effective basis on which military personnel should make decisions in ethically challenging circumstances, would selecting either approach, to the exclusion of all other options, be appropriate from the perspective of the state?

It is noteworthy that Black and King are each at pains to show that their respective model covers the ethical ground that each of the 'big three' approaches to ethics seeks to address. King describes the Help Principle as 'a correction of utilitarianism', which is also 'an expression of a virtue' as well as 'a deontological precept'. Black points to the compatibility of natural law principles of reasonableness with deontological imperatives like Kant's injunction 'not to use people as a [mere] means to an end'; he contends that the application of these principles 'are also character forming'; and he argues that natural law thinking properly accounts for 'the ethical character of the consequences we cause'. In addition to accounting for the importance of these foundational ethical intuitions, each of them also contends that their approach does not suffer from the particular shortcomings displayed by the three dominant approaches. Perhaps one of them is correct. But which one? How should the state decide? Certainly, academia is unlikely to deliver a meaningful consensus in this regard. This is another reason why the Ethical Triangulation model deliberately and explicitly eschews any claim to being a philosophically coherent *theory*. The claim is, instead, far more modest – Ethical Triangulation is a practical decision-making tool that requires decision-makers to consider the fundamental ethical intuitions that underpin the major ethical theories that shape our society.

From a practical perspective, I have two further concerns about the models that Black and King have put forward. The first is what I will call 'the problem of buy-in'. My experience of teaching ethics to military personnel in (primarily) the United States and Australia over the past 15 years has taught me that there is often an initial reluctance on the part of my audience. This is particularly so when (as is often the case) the ethics class or lecture or seminar is an element of mandatory training and education. There is a natural tendency for military personnel to feel that the ethical values and viewpoints they have gained from their upbringing and experiences are perfectly adequate to address the ethical challenges they are likely to face in the future. At the same time, those values and viewpoints are diverse, and grounded in differing personal, cultural, social, religious, and other founda-tions – increasingly so as our liberal democracies become more multicultural. This gives rise to 'the problem of buy-in': how do we get this diverse group to 'buy in' to a common model of ethical decision-making? If we present a model based on a particular philosophical position, we immediately exclude those in the military who have different foundations. If we choose a natural law theory-based model, we will have great trouble getting buy-in from military personnel who reject the idea that there is such a thing as 'a fixed human nature'. If we choose a quasi-utilitarian approach then those with religious beliefs are unlikely to come to the party. The same problem would rear its head for *any* ethical decision-making model based on a particular philosophical perspective. By contrast, a model like Ethical Triangula-tion offers a 'way in' for almost[9] all potential users.

It is important not to overlook the importance of providing the military with a common language for discussing ethics. This is vital in bridging the diverse back-grounds and perspectives that individual military personnel bring with them. I am reminded here of Michael Walzer's comment in the preface to *Just and Unjust Wars*:

Someone can always ask, "What is this morality of yours?" That is a more radical question, however, than the questioner may realize, for it excludes him not only from the comfortable world of moral agreement, but also from the wider world of agreement and disagreement, justification and criticism. The moral world of war is shared not because we arrive at the same conclusions as to whose fight is just and whose unjust, but because we acknowledge the same difficulties on the way to our conclusions, face the same problems, talk the same language.

(Walzer 2007, 19)

In a similar way, it is vital that we select ethical decision-making models for our military forces that allow for the level of 'buy-in' necessary for our soldiers to 'talk the same language' on issues of ethics.

A final concern: I wonder about the practical utility of the models put forward by Black and King in actually aiding the average soldier in decision-making. Of course, both of them believe in the utility of their models, and both are able to show how their models can be applied in practice. But they are also both deeply steeped in the philosophies underpinning their decision-making models. That is unlikely to be the case for most soldiers, even after they have been on the receiving end of their mandatory ethical training. At best, we can hope they will remember the key steps of the model themselves. For Black those are the questions 'is my intended outcome good?' followed by 'is my plan to pursue that end reasonable?'

The words 'good' and 'reasonable' in those sentences are packed full of meaning in the context of natural law theory, but if the soldier is not equipped to draw on that well then it might be the case that these questions are simply too broad to provide help in addressing specific ethical challenges. Similarly, King's Help Principle is summarized as 'Help someone if your help is worth more to them than it is to you'. While there are circumstances in which asking this question will deliver an unambiguous course of action, there are many more (including many of those that King gives as examples) where it will be necessary to extrapolate from the principle in order to reach some kind of decision. King is perfectly able to do this, but I question whether we can expect the same from the average soldier. I will leave it up to my interlocutors to offer a view as to the practical efficacy of the Ethical Triangulation model! This is another question which would benefit greatly from empirical study – which of the available ethical decision-making models do users find most useful when engaging a range of ethically challenging situations? And which of the models available shows the greatest tendency to lead to the kind of decisions that the institution wants its personnel to make?[10]

There is more I could say in response to Black and King's chapters. I am itching to discuss the question of whether the natural law puts too much weight on intention and to ask whether we should be comfortable with any theory that proffers (as King believes his approach does) a neat and clear answer to Bernard Williams' infamous 'Jim and the Indians' case. But that would be to indulge my inner philosopher, when my central point here is that, in thinking about what ethical decision-making model we select for our military forces, states should conscientiously avoid taking any particular philosophical position! So, in closing, I will instead simply thank my co-authors for their generosity of spirit in engaging in this important debate.

Notes

1 These were adopted in revised form into the plan as:

   1 Ethical knowledge and leadership
   2 Ethical Awareness and Reasoning
   3 Ethical and Moral Motivation
   4 Ethical resilience and endurance ('grit')

   (Department of Defence 2020, 2)

2 As in my previous chapter, I will use 'soldier' as shorthand for any uniformed member of the military, to include uniformed members of naval, marine, air force, and other military formations. I recognize that this is irritating for non-Army personnel, and for that I apologize!

3 Indeed, I have had the privilege of addressing these issues alongside Roger Herbert when we had the opportunity to deliver ethics training to Australia's Special Operations Command together in 2021.

4 For example Arnhart (1998).

5 Murphy continues:

   To summarize: the paradigmatic natural law view holds that (1) the natural law is given by God; (2) it is naturally authoritative over all human beings; and (3) it is naturally knowable by all human beings. Further, it holds that (4) the good is prior to the right, that (5) right action is action that responds non-defectively to the good, that (6) there are a variety of ways in which action can be defective with respect to the good, and that (7) some of these ways can be captured and formulated as general rules.(Murphy 2019)

6 See, for example, Kuhse and Singer (1985), and McMahan (2002).
7 See, for example, Richard Waswo's 'The Formation of Natural Law to Justify Colonialism, 1539–1689' (1996). Nigel Biggar, co-author with Rufus Black of *The Revival of Natural Law: Philosophical, Theological and Ethical Responses to the Finnis-Grisez School* (2000), has become controversial in academic circles in recent years through his attempt to offer a qualified defence of colonialism (see Biggar 2022 and, for an example of the reaction to his book, Owolade 2023).
8 For example, as Brent L. Pickett summarized in 2004:

> Today, natural law theory offers the most common intellectual defense for the differential treatment of gays and lesbians. Leading natural law theorists, such as John Finnis, Robert George, and Richard Duncan, have eagerly inserted themselves into debates about law and sexuality. If their arguments are sound, it is permissible to treat gays and lesbians differentially as a matter of law. Natural law theorists defend a range of policies in this regard, from antisodomy laws, to allowing persons to discriminate against gays and lesbians in employment and housing, to a fierce opposition to same-sex unions or, in their watered-down version, civil unions. In general, natural law theorists believe that it is rational for the state to 'discourage' homosexuality through such policies. Furthermore, they believe that other laws regulating sexuality are permissible and potentially beneficial, such as forbidding the sale of contraceptives to unmarried persons.(Pickett 2004, 39)

9 Some perspectives are, of course, excluded even by Ethical Triangulation. But this is acceptable – as Charles Taylor rightly points out, 'Liberalism is also a fighting creed' (Taylor 1994, 249).
10 The 'Army Intermediate Concept Measure' (AICM) tool developed by David Walker, Stephen Thoma, and James Arthur at the University of Alabama offers a way to evaluate this. Subjects are asked to respond to four ethical dilemmas, and their responses are then measured against the answers provided by a panel of experts (see Walker *et al.* 2021).

References

Arnhart, Larry. 1998. *Darwinian Natural Right: The Biological Ethics of Human Nature.* State University of New York Press.

Biggar, Nigel. 2022. *Colonialism: A Moral Reckoning.* William Collins.

Biggar, Nigel, and Rufus Black. 2000. *The Revival of Natural Law: Philosophical, Theological and Ethical Responses to the Finnis-Grisez School.* Routledge.

Department of Defence. 2020. 'Special Operations Command (SOCOMD) Ethics Plan 2020.'

Kuhse, H., and P. Singer. 1985. 'Ethics and the Handicapped Newborn Infant.' *Social Research* 52(3), 505–542.

Mauser, Bernard. 2011. 'The Ontological Foundations for Natural Law Theory and Contemporary Ethical Naturalism.' Dissertations (2009–). Paper 96. http://epublications. marquette.edu/dissertations_mu/96

McMahan, Jeff. 2002. *The Ethics of Killing: Problems at the Margins of Life.* Oxford University Press.

Murphy, Mark. 2019. 'The Natural Law Tradition in Ethics' in Edward N. Zalta (ed.) *The Stanford Encyclopedia of Philosophy.* https://plato.stanford.edu/archives/sum2019/entries/natural-law-ethics/

Owolade, Tomiwa. 2023. 'Nigel Biggar's Whitewashing of Empire.' New Statesman, March 3. www.newstatesman.com/culture/books/2023/03/nigel-biggar-whitewashing-empire

Pickett, Brent L. 2004. 'Natural Law and the Regulation of Sexuality: A Critique.' *Richmond Journal of Law and the Public Interest* 39–54.

Rachels, James. 2007. *The Elements of Moral Philosophy.* The McGraw-Hill Companies.

Taylor, Charles. 1994. 'The Politics of Recognition' in Amy Guttman (ed.) *Multiculturalism: Examining the Politics of Recognition.* Princeton University Press.

Valerie, Morkevicius. 2018. *Realist Ethics: Just War Traditions as Power Politics.* Cambridge University Press.

Walker, David I., Stephen J. Thoma, and James Arthur. 2021. 'Assessing Ethical Reasoning among Junior British Army Officers Using the Army Intermediate Concept Measure (AICM).' *Journal of Military Ethics* 20(1), 2–20.

Walzer, Michael. 2007. *Just and Unjust Wars: A Moral Argument with Historical Illustrations.* Basic Books.

Waswo, Richard. 1996. 'The Formation of Natural Law to Justify Colonialism, 1539–1689.' *New Literary History* 27(4), 743–759.

7

HERBERT RESPONSE

Roger Herbert

When my co-authors and I first settled on the unusual format for this book, my principal reservation was that the approaches to moral reasoning we were to independently submit would be so similar that our response chapters would be uninteresting, a pedantic squabble over details. While all four approaches predictably bear a family resemblance – we are all students and products of Western moral traditions – I was pleased to discover essential dissimilarities. Areas of disagreement in Chapters 2 through 5 illuminate some important arguments not only among academics but also among those we hold responsible and accountable for preparing members of the armed forces for war.

Deane-Peter Baker's Ethical Triangulation

Deane-Peter Baker kindly includes with his description of Ethical Triangulation a suggested set of criteria for evaluating his model. He writes that any ethical decision-making method for military application must be '*practical* . . . and *appropriate*' (20). My critique applies these criteria.

A *practical* decision-making method, writes Baker, 'provides an effective guide to ethically sound action in a wide range of situations' (20). A moral reasoning framework that requires a warm, dry classroom and ample time for navel-gazing is *impractical* for most military applications and certainly on a battlefield. When a wartime situation necessitates deliberate moral analysis, there's an excellent chance that things have not gone according to plan, that time is in short supply, and that the 'pucker factor' is high. Invariably (or so it seems), it's also cold, dark, and rainy. In the real world of battlefield decision-making, a moral deliberation framework that is difficult to remember or overly complex is of little practical value. When fatigue, danger, and intractable uncertainty make even simple tasks difficult,[1] soldiers will

DOI: 10.4324/9781003312925-9

justifiably jettison excessively intricate moral reasoning methodologies in favour of gut intuition and survival instincts. The moral habits they developed in their youths will just have to do, even though the moral terrain of the battlefield, as we suggest in Chapter 1, is radically novel.

Both Ethical Triangulation and the US Naval Academy's Moral Deliberation Roadmap, the model I describe in Chapter 3, deserve high marks for their practicality. But perhaps not equally high marks. Ethical Triangulation enjoys some advantages that increase – if only marginally – the likelihood that a cold, wet, scared soldier will engage in deliberate moral analysis rather than relying on instinct. First, Baker's triangulation metaphor itself is exceptionally accessible. Unlike the roadmap metaphor, which only weakly suggests a starting point for deliberation, the Ethical Triangulation metaphor hews closely with the real thing, with triangulating one's location in the physical world. Like the non-metaphorical process of triangulation, a skill with which most military personnel are familiar,[2] the process of Ethical Triangulation begins by taking three bearings – not two, six, or twelve. This point of familiarity alone, I believe, improves the chances that a soldier under duress will recall those three bearings: respect, consequences, and character.

A second feature that makes Ethical Triangulation comparatively more accessible and, therefore, more practical than the Moral Deliberation Roadmap is the number of moral factors involved: three (respect, consequences, character) rather than four (constraints, consequences, special obligations, character). Although seemingly an inconsequential advantage, every respectable poet, educator, and public speaker knows well the 'rule of three': we are far more likely to absorb and retain information delivered in clusters of three. As writing coach Roy Peter Clark (2006, 101) observes, 'In the anti-math of writing, the number three is greater than four'. Our minds, he explains, evolved to look for patterns and seek simplicity. Sets of three do both. Three is the simplest pattern (a single event is random, and a pair of events may be mere coincidence). 'The mojo of three', Clark (2006, 101) continues, 'offers a greater sense of completeness than four or more'. If Clark is correct, three-factor models should always be preferable to four-factor models, all else equal. They're easier to learn, retain, and, most importantly, access under pressure. In short, they're more practical.

But is all else, in fact, equal? Does Ethical Triangulation's parsimony come at the expense of analytical power or theoretical coherence? Ethical Triangulation's critical omission vis-à-vis the Moral Deliberation Roadmap is the Roadmap's third moral factor: 'special obligations'. This omission seems significant. Special obligations, duties we incur by virtue of roles we've accepted, relationships we've formed, or promises we've made, are constant features of our daily moral lives. They are especially salient in the lives of service members who learn to privilege their relationships with shipmates, co-pilots, or fellow occupiers of a foxhole. Indeed, special obligations are fundamental to the profession of arms. Oaths of enlistment, for example, commit members of the armed forces to run toward the sound of the guns when prudence dictates running away.

Upon closer consideration, however, Ethical Triangulation *does* account for special obligations, albeit not as a distinct moral factor as in the Moral Deliberation Roadmap. Baker's description of 'respect', the first 'moral bearing' one takes when ethically triangulating, focuses on general obligations, the duties we all incur as humans to acknowledge the dignity and worth of all other humans. But the broader role that respect seems to play in Baker's model is to account for *all* deontological considerations: general *and* special obligations. If this is right, then Ethical Triangulation does not omit special obligations. It does, however, consider these duties earlier in the deliberative sequence, before accounting for consequences. Does this difference matter? Does it matter that Ethical Triangulation only implicitly deals with special obligations?

During my tenure at the US Naval Academy, the ethics faculty debated where special obligations belonged in the deliberative process – or, better put, *when* our students should consider their special obligations. There were advocates for following Baker's lead and folding special obligations into constraints, alongside the other deontological considerations in the model. Ultimately, however, the faculty agreed that there were theoretical advantages to first considering the moral factors that apply to *all* moral agents (constraints and consequences) and then accounting for those pertaining to *particular* people (special obligations and character). Naval Academy philosopher Marcus Hedahl explains the winning argument in an email:

> Having a separate category of special obligations helps highlight the radically different role they play in morality, namely that almost always, if not always, special obligations lack the power to change the status of morally required or morally prohibited actions into morally permitted actions. You can't make it OK to rob a bank just because you promised a friend to do so; your special obligation to your friends and family cannot, by itself, override a moral constraint. . . . I can't promise to bind myself to activities that I'm morally prohibited by rights and dignity from partaking in.

An even stronger case for separating general obligations from special obligations in time and space acknowledges a potentially harmful bias that is particularly strong in those who volunteer to serve in the armed forces. Hedahl continues:

> If we recognize that our students are often going to over-emphasize the impact to their special obligations (for example, duties to their troops or peers) and under-emphasize those required by rights and dignity (for example, the duty to protect noncombatants), then there's real benefit in breaking out those categories as separate.

In the interest of honest reporting, I was among those who favoured accounting for all deontological considerations (rights, justice, *and* special obligations) before turning to consequences, *a la* Ethical Triangulation. Hedahl is right that general obligations almost always trump special obligations and that there are powerful temptations for service members to put the well-being of their comrades ahead of general moral commitments. My concern, however, is that military service often demands prioritizing our special obligations over prudential considerations, that is, our assessment of the foreseeable consequences of a given action. When I was a naval officer, I fully expected those who worked for me to bring to my attention any foreseeable adverse outcomes of my decisions. But I also expected them to salute and move out smartly if I disagreed with their consequentialist analysis. In other words, I expected my troops to prioritize their special obligation to obey the lawful orders of a higher ranking officer over their evaluation of the consequences of those orders.

In short, I can see advantages to an ethical decision-making methodology that first identifies those actions that I can, can't, or must take due to my general obligations and then, immediately on the heels of that analysis, accounts for the actions that I can, can't, or must take because of my special obligations. Only then, once I've determined which actions are morally impermissible and eliminated (or at least side-lined) them from further consideration, should I engage in a cost-benefit analysis of the foreseeable consequences of an action and its likely effects on my character.

Having said this, Ethical Triangulation would benefit from making more explicit the compound nature of 'respect'. I recommend rephrasing the foundational question that grounds this moral factor. Rather than asking, 'what ethical principles or rules apply to the situation we are considering' (27), a more fruitful question would be, 'what do I owe to other people'? To address that question, we first consider what we owe to other people because they are people, because of our general obligations to respect the great dignity, inherent worth, and inalienable rights of all humans, as humans. We can then attend to what we owe to other people because of the special obligations we incur by the roles we inhabit, relationships we've formed, or promises we've made.

The second criterion (after practicality) that Baker proposes for evaluating his model is appropriateness. An appropriate ethical decision-making methodology for a Western military is 'in alignment with the DNA of the foundational ideas of liberal democracy'. Drawing from Charles Taylor's 'perspectives of the modern self' and social contract theory, Baker conceptualizes 'two fundamental drivers . . . that define the liberal state', each of which he associates with one of the metaphorical bearings one takes when engaging in Ethical Triangulation. The first driver is the liberal state's recognition of the inherent dignity that all persons, as persons, possess, and the human rights that this assertion implies. Ethical Triangulation's initial consideration, respect, guards against rights violations. The second driver emerges from the contractarian claim that any state, liberal or otherwise, exists to protect its citizens. Ethical Triangulation's second moral factor, consequences, accounts for the imperative that state policies and battlefield decisions ultimately promote the good, understood as the safety of the polity.

Baker describes his model's first two considerations (respect and consequences) as existing in a 'virtuous tension that is foundational to a liberal democracy' (24). There is, however, a third moral factor involved in Ethical Triangulation, character, that Baker does not treat as an equal partner in triangulation. Indeed, the role assigned to the character is essentially negative. After negotiating the virtuous tension between respect and consequences, but before acting, the service member is advised to inventory her distinctive character traits to determine whether her decision has been inappropriately biased. 'Just as a strong magnet can throw off the needle of a compass', Baker (28) warns, 'the decision maker's moral identity can, if she is not careful, affect her ethical decision-making and potentially point her in an ethically problematic direction'.

Baker's rationale for assigning a relatively humble role to the character in Ethical Triangulation is the appropriateness criterion. Baker finds the idea of the state seeking to shape a soldier's character 'more than a little problematic' (20). When a military deliberately engages in character development, argues Baker, it interferes with a service member's conception of the good life. He is sympathetic, therefore, to Kasher's (2008, 139–140) contention that respect for the dignity of its members strictly limits the extent to which the military can engage in character development. Rather than shaping character, Kasher argues, military education and training should limit its normative ambitions to merely shaping behaviour, focusing on 'the principles of military ethics necessary for the effective functioning of the military force'.

Such rigorous insistence on 'state neutrality' in matters of character development is off target. Baker is right that state involvement in shaping the character of its citizens can be morally problematic and conjures examples of indoctrination policies that are anathema to liberal democratic values. But war itself is morally problematic. States train a portion of their populations to kill on a massive scale. It would, of course, be supremely unethical for a state to order its military personnel into combat without training them to fight and win. It would be equally unethical to order soldiers to kill for the state without instilling the virtues that will armour them against moral injury.[3] A Marine with a deficiency of courage, for example, will discount or disregard the lives of noncombatants the first time he hears a round crack over his head. A sailor with an excess of loyalty may save her shipmate, but the memory will haunt her that she failed to shut the hatch that would have saved her ship.

Shannon French (2017) describes the line separating the liberal just warrior from a common murderer as thin and permeable. What protects a service member from crossing this line, and has done so for millennia, is the warrior's commitment to a code of conduct – French's 'Code of the Warrior' – that strictly limits who, how, and when he may legitimately kill in war. It's not uncommon for members of the armed forces, young soldiers especially, to chafe under constraints that seem to tie their hands on the battlefield, limit their combat effectiveness, and increase the danger to them and their teammates. However, a code of behaviour that sets limits on violence is essential for the moral well-being of the soldier. Warrior codes, French argues, increase the likelihood that soldiers can endure the hell of combat and return home with their humanity intact.

The state has a stake in ensuring this outcome. Demands of reciprocity for the many sacrifices that service members and their families make obligate states to take reasonable measures to bring them home from war with body, mind, and soul intact. But the state also acts in its self-interest when it morally armours its personnel for battle. Service members who fight justly in just wars return home as whole human beings and reintegrate seamlessly into the societies that sent them to war. By contrast, soldiers who cross French's thin and permeable line return home as murderers (or rapists or torturers). They never fully reintegrate with the society that sent them to war. They pose a danger to their communities, their families, and themselves. Ultimately, they 'will find themselves condemned by the very societies they were created to serve' (French 2017, 4).

So, how best to secure a soldier's deep commitment to a code of ethics? Is Baker's minimalist approach – promoting 'principle-guided behaviour' rather than shaping character sufficient? Up to a point, yes. Military personnel are good at taking orders. However, exercising moral constraint in combat may, in some circumstances, appreciably increase the physical danger to the soldier and her team. If, for example, she must choose between the safety of noncombatants and her own safety or that of her team, she will be inclined to discount the former *unless* she possesses the strength of character to respect the principle of noncombatant immunity despite known risks and her legitimate fear. Instructing and drilling military personnel on desired behaviours will not reliably affect this. A soldier will consistently engage in principle-guided conduct through the fog and friction of war only if she genuinely values those principles and has adopted them as aspects of her identity (Osiel 1999, 23).

What, in practice, might character shaping look like? After their first day of basic training, recruits should not doubt that they have joined a community that honours particular virtues. As their professional military education progresses, service members should learn what these virtues (courage, for example) demand and what their vicious expressions (cowardice or recklessness in the case of courage) look like. Training should present opportunities to exercise virtue, and instructors should reward virtuosity while remediating viciousness. Leaders should be swift to recognize displays of virtue with medals and promotions. Vicious behaviours should prompt 'reintegrative shaming' (Osiel 1999, 35) if reform seems possible and ostracism if not.

An additional reason for the state, through its military, to engage deliberately in the character development of its soldiers is that character development is simply a fact of military life, with or without the imprimatur of the state. Typically, this informal character shaping influences the lives of young service members in profoundly positive directions. But not always. Two cases that have recently come to the general public's attention illustrate how character-shaping practices without institutional intentionality can result in substantial harm. The 'Brereton Report'[4] details how several mid-grade NCOs in Australia's storied SAS employed a shocking character-shaping practice called 'blooding'. According to the report, several SAS patrol leaders directed newly selected commandos on their first deployments to Afghanistan to execute their prisoners. Blooding was both a rite of passage and an exercise to shape character by conditioning the young commandos to the act of killing. Author David Philipps details a second case of a harmful character-shaping strategy. According to Philipps (2021), the US Navy SEAL Platoon Chief Eddie Gallagher set out to instil in Alpha Platoon, the SEAL Platoon he led, his personal disdain for moral constraint in war.

As a retired officer in the US Navy, I found Philipps' account particularly difficult to read. However, Philipps offers a positive message that has a bearing on this discussion. We know about Gallagher's alleged outrages because Craig Miller, the Platoon's Leading Petty Officer (Gallagher's immediate subordinate), knowingly put his career and reputation at risk and reported his Chief's behaviours up the chain of command. It may well be that Miller's courage to step forward and 'rat out' Gallagher – which is how some in this tight-knit community still judge Miller's actions – is that he was by nature and upbringing a man of integrity. While this is likely true, it doesn't tell the whole story.

By 2005, nearly 4 years into an armed conflict that would last a generation, it had become clear to leaders in the US Navy SEAL community that long-term exposure to high-intensity combat threatened not only the physical and psychological health of the force but also the community's moral health as well. So, along with instituting policies focused on physical and psychological readiness, Navy leadership started considering measures to morally armour the force for the long war. The initial effort took the form of a deliberate character-shaping campaign. A handpicked team of mid-grade officers and senior NCOs sequestered on a desolate island off the coast of San Diego. Their task was to craft a statement that explicitly renders the ethos of the SEAL community, calling out the character traits it honours and those it does not. After weeks of study and introspection, the team produced a creed that has stood ever since as the measure of a SEAL's moral quality.

Based on my reading of Philipps account, it is likely that Alpha Platoon's LPO was a man of strong moral convictions on the day he enlisted in the US Navy. But Miller himself tells a different story. For his decision to do the right thing, his commitment to persevere through the many obstacles set by those who sought to protect Gallagher, and the equanimity to accept the social and political consequences of his decision, Miller credits his immersion in and commitment to the SEAL Ethos. In short, Philipp's narrative is, among other things, a testament to the value of character shaping in the military.

Iain King's Quasi-Utilitarianism

One could fairly characterize Baker's Ethical Triangulation and the US Naval Academy's Moral Deliberation Roadmap as comprehensive moral frameworks condensed into field-expedient adaptations. While these models seek to simplify and streamline, they nevertheless require that military personnel consult each of the three dominant traditions in Western moral philosophy before deciding. Iain King and Rufus Black take markedly different approaches. Both authors seek to account for deontological, consequentialist, and aretaic factors, but they do so in ways that don't require the decision-maker to make three distinct judgements. This section focuses on King's approach: quasi-utilitarianism.

For King, a theory of warfighting ethics must be 'credible enough to be trusted; comprehensive enough to cover the full range of dilemmas; and coherent, so that it gives clear rather than contradictory advice' (60). His chapter pays particular attention to coherence. '[A]s soon as the ethical checklist has more than one item on it', writes King, 'there is a possibility – often a likelihood – that the different ethical considerations will provide conflicting advice, leading, once more, to incoherence' (63). Moral analysis that attends in equal measure to each of the three major traditions – like the models that Baker and I describe – promotes comprehensiveness, but risks incoherence. Analysis that yields contradictory moral guidance while lacking a method for adjudicating between competing claims could leave the decision-maker worse off than if he had relied on go-with-your-gut intuitionism. '[I]f ethical advice is incompatible with itself', he writes, 'then it is no advice at all' (62).

One obvious strategy for achieving coherence is committing to a single tradition. Undistracted by considerations of consequences, a strict adherent of deontological theory won't equivocate. She will perform her duty for the sake of duty alone, 'though the heavens fall'.[5] Likewise, utilitarianism, at least in its Benthamite formulation, offers its committed disciples a moral reasoning equation that aspires to mathematical precision. Two plus two will always result in four. If action A pumps more net good into the universe than action B, the committed utilitarian is morally obligated to take action A. However, ethical dogmatism's coherence (and comfort) is offset by its tendency to walk us into obviously bad choices. King reminds us of Kant's laboured (though coherent) axe-murderer hypothetical. Similarly, consequentialist reasoning that is not tempered by principles of human rights can justify (and has justified) abominations like chattel slavery, organ harvesting, and torture.

Comprehensiveness and coherence, King observes, are often in tension. His prescription for mitigating the tension is quasi-utilitarianism. Rather than consulting an overwhelming 'smorgasbord' (66) of moral theories that could well result in an equally overwhelming array of solutions, King suggests that service members confronting ethical challenges should begin their analysis by asking which part of the moral continuum – character, actions, or consequences – 'presents itself most clearly' and then focusing attention there (65). For King, the critical variable is the availability of information regarding foreseeable outcomes. '[W]hen choices about outcomes are clear, and information is abundant, consequence-based thinking will be appropriate' (65), and the service member would be well advised to 'suppress, moderate, modify, or redirect' (Flanagan and Rorty 1993, 1) rule-based and virtue-based considerations. When outcomes are less clear, but the moral landscape is essentially recognizable, a rule-based approach is more likely to yield moral insights. Finally, when confronted with 'wicked problems' (66) and unfamiliar situations with indeterminable consequences, the service member should rely on virtue theory, setting aside consequentialist or deontological considerations, which, given the information available, are unlikely to be productive.

King offers six vignettes to illustrate quasi-utilitarianism in action. While these were helpful, I found it instructive to apply King's method to a thornier case than those he sketches. The *Lone Survivor* incident gained public notoriety through Marcus Luttrell's best-selling book and the blockbuster movie that it inspired. The moral dilemma Luttrell describes in his account highlights several advantages of quasi-utilitarianism, but it also suggests a critical limitation.

In June 2005, Combined Joint Special Operations Task Force Afghanistan (CJSOTF-A) received intelligence that a remote village complex in Afghanistan's Korangal Valley was serving as the headquarters for a high-ranking Taliban leader, Ahmad Shah. A four-man special reconnaissance patrol, led by US Navy SEAL Lieutenant Michael Murphy, was inserted by helicopter into the rugged Hindu Kush. The team's mission was to confirm Shah's location and, if the Taliban leader was indeed in the village, to call in an airstrike.

After a gruelling infiltration, the team settled into their hide site, just as the sun was rising. Murphy's biographer describes the unwelcome encounter that followed:

> Around noon, three goat herders stumbled upon the team's concealed location. They were quickly captured, but their presence resulted in a dilemma for Murphy and the others, whose options were limited. They could kill the goat herders and compromise the mission, or they could let them go and hope they did not give away their location.
>
> *(Williams 2011, 1–2)*

Although Lieutenant Murphy was unschooled in King's quasi-utilitarianism, the debate that ensued, as told by Luttrell and Robinson (2007), suggests the rudiments of a quasi-utilitarian analysis. After rapidly brainstorming options, the team attempted to work out the likely outcomes of each course of action under consideration. The catalogue of unknowns was daunting. The team's attention then shifted – as a quasi-utilitarian analysis should when confronted with excessive uncertainty regarding outcomes – to rule-based considerations. Although the team focused its rule-based analysis on SOPs and rules of engagement, a more sophisticated rule-based approach would have yielded some solid handholds. Rule-utilitarian reasoning (a consequentialist form of rule-based reasoning) would argue that non-combatant immunity, the foundational principle of the *jus in bello* discrimination criterion, tends to result in less net suffering in wartime. Kantian ethics (a deontological form of rule-based reasoning) would reason that the goatherds had done nothing to forfeit their rights not to be harmed. The duty to recognize those rights, therefore, would have been sufficient rationale to conclude that killing the goatherds was morally impermissible.

Yet the dilemma Murphy faced was, arguably, more complex than a rules-based analysis could adjudicate. According to Luttrell and Robinson (2007), every team member understood well that in the Korangal Valley, there was no such thing as a noncombatant. In a 2013 interview with CBS correspondent Anderson Cooper, Luttrell described the unconcealable hatred in the goatherds' eyes. Despite their fervent 'no Taliban, no Taliban' assurances, the team knew that the goatherds would report their encounter with the SEALs as reliably as any military reconnaissance patrol. In short, the *Lone Survivor* case may be an example of King's wicked problems, sufficiently peculiar to warrant reliance on character, especially the virtue of empathy that King emphasizes. We learn from his biographer that Michael Murphy was a man of unimpeachable character. Ordering the execution of unarmed civilians to ensure his own safety, or even the safety of his team, would have been an act so alien to Murphy's character that it was probably never really an option.

The Taliban, as we know, rapidly responded to the goatherd's report by marshalling a large and heavily armed reaction force. The protracted firefight that ensued cost the Taliban dearly, but it annihilated Murphy's team; only Luttrell, though critically wounded, survived the firefight. Yet despite the calamitous outcome, the US military concluded that Lieutenant Murphy had made all the right moral choices that afternoon, and for his conspicuous moral and physical courage, was awarded the Medal of Honor, the US military's highest distinction for valour in combat.

Whether Murphy's moral reasoning process so closely adhered to King's prescription is debatable, but the case does illustrate how quasi-utilitarian reasoning can lead soldiers to make good ethical choices in the most stressful situations. However, with only a slight counterfactual adjustment, it's also possible to conceive quasi-utilitarian reasoning leading Murphy to the opposite conclusion. What if Murphy had decided that the likely consequences of his options were unequivocally evident? According to Luttrell and Robinson's (2007) account, one of the members of the reconnaissance patrol, Petty Officer Axelson, made precisely this argument. Releasing the goatherds, he insisted, meant four dead SEALs and mission failure, full stop. To Axelson's credit, the only difference between his prediction and how events ultimately unfolded was Luttrell's extraordinarily unlikely survival.

Had Murphy embraced Axelson's certainty, then, as a quasi-utilitarian reasoner, he would have restricted his moral purview to consequentialist considerations. So, what would a consequences-based analysis look like in this case? If the SEALs decide to execute the goatherds, the foreseeable outcome would be three dead Afghans, the possibility that the team would complete their mission and eliminate Shah, and only a remote chance that their crime would be discovered and prosecuted. If, on the other hand, Murphy orders the release of the goatherds, the foreseeable consequences include four dead SEALs, the survival of a deadly Taliban leader, and the risk of losing more Americans when CJSOTF-A predictably launches its quick reaction force (QRF) to find the missing reconnaissance patrol.[6] While determining the net good or harm of an act is always imprecise and given to disparate conclusions, it is at least arguable that Murphy – seeking to minimize 'the number of life-years lost by his unit' (66) and discounting rule-based and virtue-based considerations as quasi-utilitarianism advises – would have concluded that he was morally justified on consequentialist grounds in ordering three murders.

King claims that quasi-utilitarianism 'corrects the commonplace flaws in the traditional concept of utilitarianism' (65). My chief critique of quasi-utilitarianism is that it falls short of this ambition. Once a quasi-utilitarian reasoner can claim a high degree of certainty regarding the outcomes of an action, then quasi-utilitarianism is simply utilitarianism, with all the associated flaws and advantages of consequentialist-based reasoning. In my counterfactual application of quasi-utilitarian reasoning, for example, murder was on the table as a morally defensible option.

If King is right that 'We are entitled to accept some moral statements as facts, even though they may be inexplicable to us' (61), then surely the statement 'murder is wrong' ranks high on the list of candidates. Accounting for those moral facts and eliminating them from consideration must precede moral deliberation in wartime. Only the direst of consequences can justify setting them aside.

Rufus Black's Natural Law Revival

Like Baker and King, Rufus Black places a premium on the practicality of military ethics. He is right to do so for reasons I emphasize previously; a moral deliberation framework for military applications must be accessible even under extreme circumstances. In Chapter 5, Black makes a strong case that natural law theory can serve as the foundation for an ethical decision-making framework that meets this criterion. He argues that natural law theory, which evolved alongside, interacted with, and informed both the just war tradition and international law, can also serve as an 'eminently practical' approach to 'day-to-day decision-making' for military personnel (93).

I found it helpful to read Black's essay as a contribution – and, it seems to me, an original contribution – to a larger project. Since the early 1980s, John Finnis, Germain Grisez, and Joseph Boyle, among others, have called for a revival and reimagining of natural law theory. These scholars maintain that insights from this rich tradition can be usefully applied to public policy and problem-solving in a broad range of research fields including bioethics, environmental ethics, and legal theory. '[T]he essence of natural law or natural right theories', Black argues in this volume, 'is that they are about the application of sound reasoning' (80).

Black's chapter applies the 'Finnis-Grisez school' interpretation of natural law to military ethics. Natural law theory, he argues, is 'built around the practical reasonableness of our intended ends' (93). If one's intentions are directed towards participation in one or more of the essential goods associated with human flourishing – a set of notionally objective moral principles inherent in human nature and discoverable by human reason – then it is practically reasonable to act on those intentions. Therefore, the first step of moral reasoning using a natural law framework is 'a simple and intuitive question, "is my intended outcome good?"' (93). Understanding the intention of an act also requires an assessment of the means and how one plans to pursue the intended outcomes. This premise, for Black, implies a second question: 'are the means I am using reasonable?' (93). Means that oppose the basic goods of human well-being are judged unreasonable and, therefore, morally wrong.

A moral reasoning framework that distils such a rich and complex theoretical tradition down to two questions earns high marks for simplicity. But simplicity, while conducive to practicality, does not assure it. Indeed, Black starts his chapter by setting a high bar for practicality. Practical military ethics, he asserts, must enable service member, those in combat and in garrison, to rapidly analyse ethically complex situations and reliably make decisions that are legal, respectful of military command structures, promote organizational culture, and account for the risks of moral injury inherent in military service. Black herds these various desiderata into seven distinct tasks.

Black makes an excellent case that his model generally performs admirably vis-à-vis these seven criteria. Two of these tasks, however, warrant closer attention. I argue that a natural law basis for military ethics accomplishes one of these tasks *exceptionally* well – perhaps better than the other moral deliberation frameworks in this volume. For the other, I submit that his model underperforms.

Natural law theory enjoys a marked advantage for the final task in his list of seven. To be practical, writes Black, a moral deliberation framework for military application must 'Provide clarity on how to understand ethically the sort of "high stakes events" that are associated with causing moral injury' (79). I commend Black for introducing the prevention of moral injury as a chief criterion for evaluating the efficacy of a theory of military ethics. The two decades of war that followed the 2001 *al Qaeda* attacks in the United States have taken a severe psychological toll. But not all the unseen wounds of war are psychological. Jonathan Shay, author of the ground-breaking book *Achilles in Vietnam* (1994) that conceptually introduced moral injury, describes it as a 'soul wound inflicted by doing something that violates one's own ethics, ideals, or attachments' (Shay 2012, 57). Brett Litz and his colleagues (2009, 700) define moral injury as a loss of trust, profound guilt, or degradation of virtue that may result from 'perpetrating, bearing witness to, or learning about acts that transgress deeply held moral beliefs and expectations'.

A fundamental premise of this book is that an intentional, practical, widely embraced, and well-rehearsed moral deliberation framework can potentially armour soldiers against the worst effects of moral injury. It can accomplish this in two ways. First, an ethical deliberation framework can prevent moral injury by preventing service members from making morally injurious choices in the first place. Sometimes moral injury results from a genuinely wrongful act that the service member committed or witnessed. If a service member can recognize *ex ante* that an act is morally wrong, he can seek an alternative course of action or try to convince his leaders that they are making a mistake.

Often, however, moral injury results not from an obvious moral transgression, but rather from events in which the afflicted service member did nothing wrong. They are entirely guiltless, yet they suffer debilitating guilt. A common cause of unwarranted guilt is an exaggerated sense of a soldier's span of control over events or scope of personal responsibility. Nancy Sherman's superb treatment of moral injury in her book *Afterwar* (2015) recounts the case of Eduardo 'Lalo' Panyagua, a US Marine Corps corporal who was filling a sergeant's billet during a difficult time (2010) in a dangerous place (Helmand Province, Afghanistan). 'I was in charge', Panyagua told Sherman (2015, 61), 'and my biggest fear out there was losing any one of [my Marines]'. By the end of his deployment, he had lost three Marines. Although no one up or down Corporal Panyagua's chain of command blamed him for the casualties – war is a dangerous enterprise in which the adversary has a vote – Panyagua couldn't be convinced. Memories haunted him, leaving him 'drenched in sweat as he rewatche[d] the inner movie and relive[d] the self-rebuke' (Sherman 2015, 61). Panyagua, observes Sherman (2015, 63),

> doesn't see the inflated sense of control he inserts in constructing this picture of volitional and morally responsible agency. . . . He doesn't see that he is making the blame fit by turning an omission, for which he isn't at all culpable, into a transgression that will hold him blameworthy.

Cases like Corporal Panyagua's, in which moral injury results from a false or exaggerated sense of moral culpability, suggest another role for military ethics in protecting service members: clarifying the boundaries of moral responsibility. In this regard, Black's natural law approach to moral reasoning is at its best. Natural law theory, as framed by Finnis-Grisez scholars, emphasizes intention. We are morally culpable only for the *intended outcomes* of our actions. We bear no moral responsibility for the *unintended consequences*. The degree of culpability is variable if harmful outcomes are *foreseeable* though unintended. The principle of double effect, another important idea that evolved from the natural law tradition, asserts that if our foreseeable but unintended harm is unavoidable in pursuit of the good intention *and* the foreseeable but unintended harm that results is offset by the good the act produces, the soldier should not be held morally blameworthy by others or herself.

In Corporal Panyagua's case, it seems probable that even if he had fully internalized, intellectually and emotionally, these distinctions regarding intentionality, he would have been saddened but not morally injured by the foreseeable but unintended combat deaths of his Marines. Furthermore, if the Corporal's fellow

Marines – his superiors, peers, and subordinates – had fully internalized, intellectually and emotionally, the distinctions that natural law theory makes regarding intentions, they would have been better prepared to address Panyagua's unwarranted guilt. Of course, we can't be sure this would have made a difference. My sense, however, is that any measure – *ex ante* or *post hoc* – that clarified Corporal Panyagua's understanding of the boundary of his moral responsibility would have at least attenuated the extent and duration of his suffering.

I'll turn now to the task for which Black's natural law approach to moral reasoning underperforms, in my estimation, the other models proposed in this text. Black correctly asserts that a practical moral reasoning framework must 'Enable military personnel engaged in armed conflict to make very rapid, reliable, and consistent ethical decisions under conditions of extreme pressure' (78). Presumably, a deliberative framework grounded in natural law theory should excel at this task, especially in ensuring consistency. Indeed, the vision of the original natural law theorists, most notably Thomas Aquinas, was a moral theory that would inform moral reasoning regardless of time, place, culture, or circumstance. Rationally deriving values, norms, and obligations from the basic goods of human flourishing should yield moral injunctions and prohibitions that would apply equally to all humans.

My first concern is that the reliability and consistency of a natural law approach depend on a sophisticated understanding of natural law's conceptualization of what is 'good' and what is 'reasonable'. Black does an admirable job explicating these concepts, but I, for one, do not find them uncomplicated. If a conference of natural law theorists were given the same scenario and tasked with recommending a course of action based on Black's grounding questions, 'is my intended outcome good?' and 'are the means I am using reasonable?' it's reasonable to imagine that there would be a significant degree of consistency in courses of action selected and in the justifications for those selections. But as Black observes, 'you can't train all frontline military personnel to be ethicists' (86). Unless soldiers are immersed in the literature and language of natural law, 'goodness' and 'reasonableness' are insufficiently cogent concepts to ensure consistent decision-making outcomes.

My second critique regarding the reliability and consistency of Black's model is one that we raise in the introduction to this volume. Our book starts with a case study in which a patrol leader must choose between two awful options. If the patrol leader decides to hunker down and wait for a fire mission to eliminate the machine gun position that has your team pinned down, one of his soldiers will very likely bleed out from wounds sustained from that gun. Alternatively, he can order his sniper to target the individuals who are clearly ferrying ammunition to the gunner. Ordinarily, this would be an easy decision. In this case, however, the ammo haulers are children.

As the reader may recall, we introduce this case study because it illustrates the limits of both legal analysis and just war theory. Both courses of action are legal. Both courses of action fulfilled the *jus in bello* criteria of discrimination, proportionality, and necessity. As this case demonstrates, the law and the just war tradition can advise soldiers that a given action is legal or illegal, just or unjust. The task of working through a no-kidding moral dilemma, however, is beyond the scope of these institutions.

If the patrol leader in this scenario were to reach for the ethical decision-making method that Black describes, I think he would be similarly disappointed. Having determined that for either option – hunker down or kill the children – his intentions are good, and his means are reasonable (as natural law theory understands goodness and reasonableness), he is left with the same question that prompted his moral deliberation: what is the right thing to do? Which of the two good-intentioned actions *ought* the patrol leader pursue? Like the *jus in bello* principles, Black's criteria are excellent for determining whether a given course of action is morally permissible. However, they are not as powerful for working through moral dilemmas, a regrettably common feature of military decision-making.

Conclusion

I stepped down from my faculty position at the US Naval Academy in the summer of 2021 and, therefore, no longer influence the core ethics course that I formerly directed. However, if I still occupied that position, I would have suggested several adjustments to our curriculum based on the excellent work of my co-authors in this volume.

First, based on Deane-Peter Baker's contributions in Chapter 2, I would take another crack at adjusting our model. I'm more convinced now that decision-makers applying the Moral Deliberation Roadmap should consider special obligations – the moral obligations we incur due to our roles, relationships, or promises – before weighing consequences. I'm convinced that Ethical Triangulation has this right. I would also draw from Baker's model, his emphasis on *phronesis* or practical wisdom. After working through either the Moral Deliberation Roadmap or Ethical Triangulation, the soldier is left with a shortlist of morally permissible actions. So, now what? Baker stresses that the practical wisdom to turn moral deliberation into moral action relies on the decision-maker's experience-informed judgement. Students should understand that developing practical wisdom is a lifelong endeavour. If approached intentionally, practical wisdom is a virtue that can be refined and strengthened with every moral choice we make.

Drawing on Iain King's quasi-utilitarianism, I would suggest revising how we counsel our students to adjudicate between the Moral Deliberation Roadmap's four moral factors. As outlined in Chapter 3, the Roadmap is a sequential framework that will almost always flow in one direction. We instruct our students first to subject considered actions to deontological considerations and then apply consequentialist reasoning only to those actions that 'make the cut' as morally permissible. When I taught our core ethics course at the US Naval Academy, I advised my students that consequences trump constraints *only* in extreme cases. The final two moral factors of the Moral Deliberation Roadmap, special obligations, and character, serve essentially as tiebreakers. King's approach to adjudication between moral factors, with its emphasis on the service member's assessment of the information environment, has caused me to reconsider my thoughts on this. King offers us an intuitively appealing rule of thumb that appropriately elevates the importance of consequences and virtue.

Finally, I'm intrigued by Rufus Black's natural law revival and the direction that the body of scholarship may be heading. When the US Naval Academy revised its curriculum, we essentially demoted natural law. In the original version of our curriculum, we devoted an entire week, and a chapter in our textbook, to exploring natural law. In the current curriculum, we acknowledge natural law's contribution to the evolution of the principle of double effect and then quickly move on to other topics. This demotion was not universally embraced by the faculty. After reading Black's chapter, I wonder if we cut too deeply. As the Finnis-Grisez interpretation of natural law that Black outlines continues producing insights, the Curriculum Committee for the Naval Academy's core ethics course may be well advised to consider a natural law revival in Annapolis.

Notes

1 Here I'm paraphrasing the Prussian general and military theorist Carl von Clausewitz who observes, 'Everything in War is very simple. But the simplest thing is difficult' (Clausewitz 1976, 119–121).
2 Triangulation is an essential tool for land navigation, coastal piloting, close air support, artillery fire, or any other military skill requiring geospatial precision.
3 I credit moral theorist David Whetham for the 'moral armouring' metaphor. See Baker *et al.* 2023.
4 The unofficial title of this report comes from the lead investigator for the inquiry, Major General Paul Brereton, Assistant Inspector-General of the Australian Defence Force. The official title is the 'Report of Inquiry under Division 4A of Part 4 of the Inspector-General of the Australian Defence Force Regulation 2016 into Questions of Unlawful Conduct Concerning the Special Operations Task Group in Afghanistan'.
5 *'Fiat iustitia ruat caelum'* (let justice be done though the heavens fall) is attributed (and misattributed) to multiple authors.
6 Operation Red Wings demonstrates the need to take seriously the risk to rescuers when deciding to call for help. CJSOTF-A dispatched a Chinook helicopter with an embarked QRF to come to the aid of Murphy's team. When it was shot down, killing all 16 special operators on board, 28 June 2005 became the deadliest day in Naval Special Warfare history up to that point.

References

Baker, Deane-Peter, Roger Herbert, and David Whetham. 2023. *The Ethics of Special Ops: Raids, Recoveries, Reconnaissance, and Rebels.* Cambridge University Press.
Clark, Roy Peter. 2006. *Writing Tools: 50 Essential Strategies for Every Writer.* Little, Brown, and Company.
Clausewitz, C. 1976. *On War.* Princeton University Press.
Flanagan, Owen, and Amélie Oksenberg Rorty. 1993. *Identity, Character, and Morality: Essays in Moral Psychology.* MIT Press.

French, Shannon E. 2017. *The Code of the Warrior: Exploring Warrior Values Past and Present*. Rowman & Littlefield Publishers.

Kasher, Asa. 2008. 'Teaching and Training Military Ethics: An Israeli Experience' in Paul Robinson and Nigel de Lee (eds) *Ethics Education in the Military*. Routledge, 147–160.

Litz, Brett T., Nathan Stein, Eileen Delaney, Leslie Lebowitz, William P. Nash, Caroline Silva, and Shira Maguen. 2009. 'Moral Injury and Moral Repair in War Veterans: A Preliminary Model and Intervention Strategy.' *Clinical Psychology Review* 29(8), 695–706.

Luttrell, Marcus, and Patrick Robinson. 2007. *Lone Survivor: The Eyewitness Account of Operation Redwing and the Lost Heroes of SEAL Team 10*. Little, Brown, and Company.

Osiel, Mark J. 1999. *Obeying Orders: Atrocity, Military Discipline, and the Law of War*. Routledge.

Philipps, David. 2021. *Alpha: Eddie Gallagher and the War for the Soul of the Navy SEALs*. Crown.

Shay, Jonathan. 1994. *Achilles in Vietnam: Combat Trauma and the Undoing of Character*. Scribner.

Shay, Jonathan. 2012. 'Moral Injury.' *Intertexts* 16(1), 57–66.

Sherman, Nancy. 2015. *Afterwar: Healing the Moral Wounds of Our Soldiers*. Oxford University Press.

Williams, Gary L. 2011. *SEAL of Honor: Operation Red Wings and the Life of Lt. Michael P. Murphy, USN*. Naval Institute Press.

8

KING RESPONSE

Iain King

Soldiers fight together, but each one fights their own war. Experiences, viewpoints, and recollections will always differ.

Writing this book, it's become clear the same can be said of war theorists. All four co-authors have all tried to derive the best advice for frontline military personnel; we have all considered similar scenarios; and all drawn on the same foundational metrics of right and wrong – consequences, rules, and virtues. Yet four very distinct positions have emerged.

Here are my reflections on those three other positions, with a conclusion on what they may mean for my own quasi-utilitarian approach.

Ethical Triangulation – Dean-Peter Baker

Whenever moral philosophy lessons begin for a cohort of university students new to the topic, one character is sure to be present in every lecture hall: the 'freshman relativist'. This is the young student who counters every suggestion for moral norms with the retort, 'That's just your opinion – how can you say that opinion is better than anyone else's?'

Deane-Peter Baker has conjured this individual and pre-empted the freshman relativist's attack on any form of ethics in military affairs. After all, how can Western liberal democracies, judiciously neutral between different ways people should live their lives, ever justify force – including the threat of violence – to compel anyone to do anything?

The answer, as Baker correctly states, quoting Charles Taylor, is that 'Liberalism is a fighting creed'. Democracies need to defend themselves and their values, which means using force to protect people both within and outside their borders. Guarding the three neutralities – of justification, intent, and outcome – for different

DOI: 10.4324/9781003312925-10

ideas of how people should live their lives requires non-neutrality towards those who pose a threat. It is an important starting point.

And by opening there, Baker's paper highlights a key deficiency in the most well-established frameworks to justify the use of force. The just war tradition and the principles of *jus in bello* are aimed squarely at top-level policy makers – the Prime Ministers and Presidents making the most fundamental decisions on war. Just war theory offers little practical advice to the actual war-fighter.

Baker's initial scenario, of soldiers passing a local woman being harangued for witchcraft, when they are en route to confront a mid-level insurgent, is all too real. What would *you* do? The scenario presents a practical challenge as much as an ethical one, highlighting how moral philosophy is most valuable when it is intertwined with reality.

Most officers confronted with this situation would, first of all, try to gather more information. The interpreter has already given a useful assessment. Is there something he's missed? Perhaps there is more to discover, leading to a secret solution which allows the dilemma to be dodged. We all want to save the woman *and* deal with the insurgent so that the problem dissolves away. But if the answer comes back that, really, a choice has to be made, then decent ethical decision-making will be essential.

So, Baker's chapter fulfils a central need: war fighters do need a model to guide them through ethical quandaries. He is right to show that such a decision-making system is required for junior and mid-level officers, not just those deciding on the fundamental decisions about whether to go to war at all. And he is right about the need for an over-arching justification for liberal democracies to engage in military activity.

His main argument, the 'ethical triangulation model', advises people to judge a difficult decision with ethical aspects to it from three viewpoints, and he highlights the trio of ethical models which dominate Western philosophy. If the three viewpoints agree that a single course of action is best, then it should be taken. So far, so good.

The difficulty with this approach lies in the detail. Baker lists the drawbacks of each approach. His critiques of virtue ethics (it usually offers too little specific advice) and of rule-based ethics (often too rigid, and too indifferent to the consequences) are both fair and appropriate. But his takedown of consequentialism is off beam.

Unfairly, Baker conflates consequentialism with Utilitarianism and then lists the faults of that approach. But consequentialism can be adjusted to avoid most of those problems. For example, he cites ethical egoism, but this is only a problem for those who apply the method wrongly. Similarly, motives can be factored in – it's the outcome being sought which counts, and if that departs from what actually happens, then we have a misjudgement about not values, but situational understanding. His only valid critique of consequentialism lies in the practical difficulty of getting accurate and timely information about what might happen. To be clear, it is a huge deficiency and provides reason enough for us to avoid a reliance on a consequence-based approach – especially in wartime, when an enemy is secretive, events move fast, and there is a bonus to causing a surprise. But by emphasising the much lesser flaws of Utilitarianism too, he biases his approach unfairly against judging options by their effects.

This bias against effects-based approaches leads Baker to argue that rules – 'Respect, Constraints and Commitments' – should come first. His justification for this carried weight: rules represent the accumulated ethical wisdom of the ages, and they can offer a good guide for achieving the best consequences. But is this really enough of a reason to justify putting rule-following before seeking good outcomes? We get a better guide for achieving the best consequences by looking at the consequences themselves. Of course, when the consequences are unknown or unknowable, we can default to rules. But when we know what can happen, surely that's the best place to start our thinking – no?

More importantly, even though rules do represent the accumulated wisdom of the ages, exactly what sort of wisdom is that? There are two possibilities. First, wisdom can come through a savvy understanding about the wrinkles of human nature. This is about knowing what tends to happen in response to a certain action or activity – for example, that most people will fight back when attacked, and that they will accept huge costs to preserve their pride. This wisdom about information is a short-cut for determining likely outcomes, so consequences can be judged accurately. 'Offer your enemy a way to retreat', for instance, is the consolidated conclusion of research into what has generated the best outcomes before. And with this wisdom, we need to be careful, because it is all based on the past. Ethical advice is needed precisely because the situation being faced is new or unusual. By definition, wisdom based on what has happened before can offer only an incomplete guide.

The other sort of wisdom is about rules which have an intrinsic ethical value: lying is inherently bad; keeping promises is fundamentally good, and so on. The values liberal democracies hold dear, such as the right to vote and the right to free speech, need to be protected strongly, even when they might lead to unwelcome outcomes. Also, by making clear our determination to uphold these fundamentals, we can frame the range of consequences which may emerge. But are all of these precepts really so fundamental that they should be where we start? Surely, the natural place to start is the situation, and from an understanding of that, the obvious next step is the ways in which things may develop. Only when we know what we're dealing with can we examine which principles may apply.

There is a further difficulty with the 'Ethical Triangulation Model', and it comes when the different viewpoints disagree. If consequences point one way, rules point another, and virtue ethics somewhere else, what should you do? Baker suggests 'It becomes a matter of wisdom'. But is that enough? If the querent had wisdom already, then she wouldn't need the triangulation model, and if she didn't have wisdom, then she wouldn't be able to use it anyway. This recourse to 'wisdom' seems like an evasion – much like a metrologist refusing to offer a forecast, and instead telling people to 'use their wisdom' to determine what the weather will be themselves.

Baker justifies this non-answer by suggesting there is a limit to the precision which can be given. Perhaps, and sometimes the choices on offer are only marginally different, whichever way they are measured. But by failing to even attempt a reconciliation between outcomes, rules and virtues (and Baker suggests they may be irreconcilable), Baker is on the same bench as the freshman relativist, arguing no sides can be taken, even when they can, they should, and they must. And, at risk of amplifying this point too much, war makes the need to decide acute: the stakes are high, and reserves of wisdom are often too low to be relied upon. Judges, city planners, clinicians, epidemiologists, and people in many other professions have developed ways to trade-off considerations between outcomes, rules, and virtues. Military advisors should too; it can't all be delegated to 'wisdom'.

Baker cites a range of sources who advocate his approach or suggest something close to it. Several Western staff colleges teach a version of the ethical triangulation model. The basic presumption behind it, of three distinct and valid perspectives which can be applied to ethical conundrums, is well founded. And Baker is correct that the just war tradition is indeed an amalgam of consequence, rules-based and virtue approaches to ethics: the six commonly cited conditions for a 'just war' draw on two of each.[1]

However, there is a significant difference: the just war tradition requires all six conditions to be met to legitimize conflict – if one or more is not, then a war is determined to be 'Not Just'. Hence, through a bias towards the status quo, the just war theory can adjudicate on every situation – no resort to 'wisdom' is required, and there is no need to balance one consideration against another. The tradition provides a system for testing which is *always* decisive. Those decisions may be contentious,[2] but the point is they are clear: 'just war' *always* offers conditions to determine what should be done, while 'Ethical Triangulation' sometimes shrugs and defers to 'wisdom'.

The predominance of ethical triangulation in one form or another allows one critique of Baker's model to be dispelled: his model is not naive. Baker cites Ordiway, who suggested the three-way analysis was somehow unrealistic, because most military personnel when confronted by a significant dilemma will think about their career and reputation, and perhaps feel strong emotions which guide them in a particular way. Ordiway is only half-right: some people will react like that, but many will respond more appropriately, especially if they have been trained to follow Baker's model. In other words, Ordiway is highlighting the problem which Ethical Triangulation tackles, if taught correctly. Any weight in Ordiway's assertion is merely ammunition for Baker to fire back. Rather than being holed by Ordiway's critique, Baker's model is reinforced by it.

Towards the end of his article, Baker is circumspect about his model: it does not promise 'a theory of ethics', he writes, merely a means for practical decision-making. In this context, the approach is absolutely valid and likely to be very useful when taught widely. It offers a decent map of the moral terrain. Those with the ethical compass Baker calls 'wisdom' will find the map helps them navigate; those without may be less lost than they were, even if they still don't know which way to go. Ethical Triangulation is a pointer towards the right direction, even if it does not provide the final destination.

Naturally, this response has focused on the negative aspects of Deane-Peter Baker's fine work. That is partly because the strongest arguments for his approach have been set out eloquently by Baker himself. And where there are negative points to be made, they are primarily about what his model misses out, not what it contains. The world would be a much better place if all military personnel were schooled in Ethical Triangulation; if only they could be given the wisdom to triangulate when the three corners of the model point in different directions. Perhaps that is for the next class, or the next paper, and they need to be taught Ethical Triangulation first.

One person who definitely should attend all the lessons is the freshman relativist. Deane-Peter Baker has given him a clear, compelling and comprehensive explanation of why some opinions definitely *are* better than others. By presenting Ethical Triangulation as a foundation for military decision-making, Baker is right, and the freshman is wrong. The relativist has been answered, and Baker's answer is fundamentally a decent one.

Moral Deliberation Roadmap – Roger Herbert

One of the most common grumbles against the compulsory teaching of mathematics, often made by middle-aged people who have settled into working life is that they can see no practical benefit from their high-school maths. And it is true that abstract notions about the internal angles of a triangle, quadratic equations, calculus, and so forth, have little direct value to most people, most of the time.

The same complaint is also made about philosophy, loaded with evidence from modern teaching of the topic which goes from the abstract to the absurd. For proof that lying is sometimes appropriate, students are told to imagine an axe murderer at their door, asking them where is the friend whom he wants to kill. For evidence that causing death is not just about numbers, we are given the example of a botanist in a very remote rainforest, who is invited to shoot someone to save nine others. And to show all sorts of things, we are given the tired scenario of a runaway trolley barrelling towards a split in the tracks. Has anyone ever really faced any of these situations? I suspect no one has – ever. This detachment from reality affects moral philosophy deeply: it means too much is hypothetical, and so less useful.

Roger Herbert has more than dodged this deficiency. He has made a practical, real-life dilemma the very centre of his argument. The tension between the captains of Riverine Command Boats 802 and 805 is all too imaginable. And amid the jarring egos, Herbert adds a dash of spilt engine oil, difficult navigation, flawed command structure, confused information, and riptide of cultural issues – from the fighting ethos of the US Navy to the clash between the Iranian and American military forces.

Too often, philosophers have bagged up these and similar factors and put them to one side to seek out the kernel of the problem. But this misses the point: the kernel is the complexity itself. It is a conceit to pretend we can separate these things, or even ignore some of them. Like life, war is incredibly messy. A practical guide to making good choices needs to help people wade through the mess and to keep wading in the right direction despite raging emotions. It is no use trying to keep our feet dry by dreaming up stepping stones along the way.

Similarly, the people most likely to need advice for these real-life messy choices are not the top bureaucrats, policy makers, or professors. Some of those will already be familiar with the tenets of Kant, Mill, and Aristotle, and many will have higher degrees in public policy making or similar topics, giving them a framework for deliberations. It is the warrior on the front line who needs advice the most – they are likely to be young and to have less life experience to draw on. They are also the most likely to face a truly testing scenario, like the Farsi Island incident of January 2016.

Herbert offers three questions to guide decision-making: Do I recognize this moral terrain? How much time do I have? Do I already know the right thing to do? The first and third of these are similar: both assume the actor asking the questions can draw upon a worthy reserve of ethical understanding. Hence, the advice being offered is targeted at a middle bracket of warriors who already have significant moral wisdom, but still need straightforward prompts. This is reasonable – after all, we would hope almost all of our frontline service personnel were in that category. But it does mean Herbert's method is more about summoning pre-existing intuitions about right and wrong, which are assumed to be correct, than about explaining what should actually be done.

Nevertheless, those prompts have to potential to be very useful; it's possible they've saved lives already. There is a worldliness to reminding us to use all the time we have available (but no more), and it is especially appropriate for warriors making tough decisions under extreme stress. Also, as Herbert explains, most people know what they really should do most of the time. It is just a question of encouraging them to do it.

Here, though, we stumble across a presumption which should not be automatic, and it concerns integrity. The quote from Admiral McRaven's 2013 Naval Academy Lecture implies that personal integrity should be at the centre of an ethical worldview, and Herbert seems to concur. The notion offered is a little simplistic – as though integrity was merely about 'sacrificing something we desire', and as though tests of integrity were just about not stealing from the cookie jar when no one is looking.

But integrity is more complicated. For a start, it is an individualistic concept and one which is sometimes at odds with how we need to behave in groups, especially tightly knit military units. The Officer in Charge (OIC) in the Farsi Island incident put his integrity aside: he surrendered and thus departed from his own strict 'never surrender' code, to do what was right for his crew. The team comes first, and personal integrity must come second to that. To put one's own personal integrity before the needs of one's military unit is close to vanity.

Frequently in war, we need to 'take one for the team'. From time to time, we need to do bad things: on those occasions when the behaviour of others means any better actions would lead to a worse outcome.[3] When something much greater is at stake, we should sacrifice our personal integrity to confront the root cause of the problem – we should not wash our hands when a crisis strikes, but get them dirty as we grapple with what is broken. Integrity matters, but it should not be over-rated, because other things matter too, and some of them matter more.

At the heart of Herbert's Moral Deliberation Roadmap are his four moral factors which are 'sometimes in conflict but always in play' in ethical decisions: constraints, consequences, special obligations, and character. This is essentially correct – although special obligations could be a subset of constraints, it is reasonable to view them separately, too. Herbert is right that there are no other factors with comparable ethical significance, and it is fair to regard constraints and consequences first to illuminate general moral obligations, relegating special obligations and character to a lower tier.

In placing constraints first, though, Herbert may be taking a legalistic approach. Legal permission does not make something right: those depraved states which commit mass murder usually enact laws to legalize their atrocities. Similarly, intervening to stop genocide – or even a single unjust killing – usually contravenes several boundaries and constraints. Nonetheless, it can be the right thing to do.

By quoting Skerker, who bases his three-part test on Kant, the author is ruling out some decent deeds which really should be allowed. Surprise birthday parties, for example, fail the universalization test (because it would not be possible to maintain the secrecy of a celebration on a known date if everyone did them). The same is true of paying off one's credit card every month – the finance companies rely on a few people to pay interest to sustain their systems. Are surprise birthdays and avoiding debt really so bad?

Similarly, although there is benign intent in not treating people as 'mere means', war usually requires that an enemy is defeated, and we must treat enemy combatants as means to that goal. Of course, their rights need to be respected, and the rubrics of international law upheld; but for a service member under fire and returning it, treating the enemy as anything other than a target can seem a little academic. Herbert is correct that rights are often waived in war, just as Joe Frazier waived his right not to be punched by stepping into the boxing ring. Because most participants in combat voluntarily surrender very major rights when they put on their uniform and take the oath, rights are not as useful a guide for moral deliberations in war as they are in most other human affairs.

Herbert does well by highlighting special obligations, and his paper provides a neat summary of so many three-way dilemmas – between our own voluntary obligations, our compulsory obligations of solidarity to those closest to us, and our more general obligations to others. But he does equally well by downplaying their role in ethical decision-making. To give special obligations any role at all, we must understand the basis for giving certain people preferential treatment. Where there is a basis, it lies in the good consequences it can generate, and in conventions based on generating those welcome outcomes: it is because we know more about those closest to us that we are more able to help them well, and assigning people to bubbles of mutual responsibility eases coordination of help to people in need. But both of these are 'tiebreakers', as Herbert rightly says. Indeed, the temptation is usually to overplay the importance of these special considerations. In other words, putting special obligations at the centre of our decisions is a form of Groupthink, in the most literal sense of the term. Usually, special obligations need to be factored out, not factored in.

Finally, character, which Herbert describes correctly as a 'Big sack of habits'. Since most of our actions are instinctive rather than deliberate, habits play a huge role in what we do, and so character deserves a very large place in any rounded exposition of ethics. This is especially true on the battlefield, where time is precious and emotions are high. But Herbert's analysis of character is based on assumptions about psychology which are not automatic. Do we really affect our character every time we choose, as Herbert claims? If we value bravery over caution, for example, does that really mean we should tilt the scales that way when we are judging what to do? There is some evidence pointing the other way: that one brave act satisfies our self-image, giving us the confidence to be more cautious next time.[4] Surely the subject matter itself should dictate; deciding whether the medalled costume of bravery looks better on us than the string vest of caution is just another vanity. Like integrity, it is a tie-breaker – doing what is right must come first, and our own moral wardrobe a distant second.

It is this very common misunderstanding of psychology which gives rise to moral injury, very appropriately cited by Herbert. Moral injury is real, and a certain sort of remorse can inflict terrible and lingering mental harm on those who have been involved in war, both fighters and non-combatants. As Herbert implies, it is far better to prevent it in the first place than remedy it after the fact. But the way to prevent it is not for people to try to do what's brave, humble, kind, clever, or whatever other virtue they imagine themselves as being. It is best prevented by people doing what is right, alongside a realistic understanding of the limits of their own volition. People should not blame themselves for things about which they could make no better decision. Failure to understand this is at the root of moral injury, not whether war fighters under fire were brave, clever, or humble.

All in all, Herbert has presented a strong and well-structured set of guidelines. It is memorable even when ordnance is whizzing overhead, and the enemy closing in. It is battle-hardened. Although this riposte has naturally focussed on its flaws, those are generally the exceptions to an otherwise worthy and useful approach. Herbert provides straightforward advice suitable to most service members most of the time.

Perhaps best of all, it answers the critique from those who grumble at their compulsory mathematics lessons. Herbert's method is based on reality, grapples with the messiness of real life, and is tested by recent historical cases – most notably, the stand-off near Farsi Island in 2016.

Like Herbert, I admire the OIC's decision – a choice made in extremely testing circumstances. When hostile elements were preparing to come aboard his Riverine Command Boats, he made the right call.

Herbert's approach would have been useful to him. And, unlike the quadratic equations and calculus of high school, it should be genuinely useful to the next generation of war fighters, too.

Natural Law – Rufus Black

To a soldier under fire, the urge to protect themselves and their comrades is over-whelming. Their every action will be directed to staying alive. There is no time for ethical nuance: whoever is sending munitions towards them is wrong.

To some, this may seem like a reason to dismiss the clever, seasoned, and care-fully thought-through work of Rufus Black, and his case for a 'Natural Law' basis for military ethics. After all, for those facing lethal peril, ethical impulses are a distant luxury. But, in fact, the two situations are far more closely related than they first seem.

For a start, Black argues that we should act ethically because it confers a stra-tegic advantage, which should be welcome to the soldier under fire. It would be wrong to regard a military benefit as the primary reason to follow moral advice. Several non-ethical actions (e.g. secretly killing prisoners of war) may also con-fer a strategic edge in some cases; that advantage certainly does not make them right. Nevertheless, by highlighting the point, Black is showing that ethics are not abstract or distant, but present and important.

Black is clear to set the limits. He takes the traditional approach of decrying the naturalistic fallacy – that we shouldn't derive notions of value from how things are. This may be unambitious. After all, our instincts to survive and reproduce, honed over many generations, guide us with pleasure and pain to do certain things. Evolution has conditioned us to ascribe value to those things which help us and our relatives. Black's 'Natural law' approach might not want to do it, but nature itself has taught us to attach moral sentiments to specific situations – to get an 'ought' from an 'is' (or, more accurately, from our ancestors' 'was').[5]

Black's chapter draws much from common instincts and intuitions. That gives it an appeal – especially to the soldier under fire, for whom these most basic drives will be dominating their thoughts and actions.

Drawing on Grotius, Black's 'Natural Law' is about sound practical reasoning, and he cites the importance of consistency, alongside ethical logic and the need for simple and consistent rules. Of course, this is essential: a set of advice which provides contradictory instructions cannot offer guidance. But his approach goes further because he provides a means by which possible advice can be ordered, or ranked, thereby assuring it will be consistent. Black argues that we should focus our assessment of actions on their foreseeable, intended consequences (and he offers a process for dealing with foreseeable but unintended side effects – more on that later).

He justifies this by linking it with Aristotle's work on practical reasoning, and the ancient Greek's observation that people seek human flourishing, and 'What is humanly worthwhile'; it is unreasonable to pursue anything else, says Aristotle, and Black concurs. But since almost everyone's day-to-day activity is driven by their pursuit of foreseeable and intended outcomes (and that includes most people in war, too), this Aristotelian association with human flourishing is superfluous.

It does, though, lead to an interesting detour, and this is about intention. Black cites an example of self-defence which involves killing the assailant, and he justifies this as reasonable because the assailant's death was unintended. He backs this up with a secondary check: would the self-defender's plan have been successful if both he and the assailant had survived? If the answer is yes, killing for self-defence is justified. Hence, Black has taken the route marked out by the principle of double effect. He offers a further justification by citing the inappropriate feelings of guilt often felt by military personnel who cause the death of innocent people.

This approach is rooted in the delusion that we should only desire purely good things. Life – and especially life in war – comes with compromises and difficult choices. Purely good options are rare; most good outcomes come with bad side effects. Departures from the status quo tend to cause at least mild harm to those who benefit from how things are now. When we decide to act, we are choosing a new state over the current one. Comparing two or more mixed options can be hard enough as it is; there is no need to overlay it with a pretence that one's own actions must be purer than the situation we face. Labelling the good aspects of our choice intended and the other unintended is a sham: when we know they come together, we are choosing them both. Would the person defending themselves have preferred their assailant to have survived? Of course. But killing was still intended, whether it was in self-defence or not.

Black offers a further justification for making intentions central to his approach when he cites examples of things going wrong, such as a bomb being dropped which kills innocent civilians. It is probably a crime if the killing was deliberate; it is 'only' a tragic mistake if the civilian deaths were accidental. But if it was an accident, further questions need to be asked about why such a terrible mistake was made. Where was the error? Could it happen again? How much risk should we tolerate?

Practical realities soon take us back to choosing between sets of options, and when our information is limited or unreliable, we may resort to rules or other guidance to fill the gaps. The difference between identical outcomes, where one is regarded as a tragedy and the other as a crime, is about where the error lies: in wrong information or in nasty preferences. Excusing bad things because they were done with good intentions may risk inviting further fatal mistakes to be made (even though it may also relieve those suffering unduly from a sense of moral injury).

Linked to this, Black invites the protagonists for whom his advice is proffered to ask themselves whether they have good intentions. Specifically, he invites the questions: 'Is my intended outcome good?', and 'For the people I am trying to help, will my intended actions make their life better?' These are fair questions, and they are appropriate, because an honest answer to them will reveal vital data. But it is naive to presume they will be answered in a fair way. When was the last time you met someone sober who admitted they had bad intentions? Even someone driven by revenge is likely to regard themselves as well intentioned – they're 'teaching her a lesson' perhaps, or 'making sure he will never do that again'. And when you add the caveat, 'For the people I am trying to help', almost anything goes. Even the most craven, imperialistic and unprovoked war can be justified by such a maxim – just believe the conflict will end quickly, or that it will make people love their nation more. In the wrong hands – especially those with trigger happy fingers – Black's maxim, taken at face value, is too self-trusting, and so too licentious. Black has identified a fundamentally important point – maybe this outcome test really should be the primary challenge to ethically questionable behaviour. But to make it work in practice, it needs to be handled differently. Perhaps the question needs to be asked indirectly, or by someone other than the protagonist.

There is a related problem with Black's resort to the word 'reasonableness', when talking of the need for proportionality. Here, though, he may simply have hit on an intractable problem, and merely highlighted that there is no perfect solution. Correctly, Black pays only lip service to the supposed distinction between ends and essential means.[6] His test is whether the means are 'reasonable', and he instructs military personnel: 'In taking action against the enemy, be proportionate'. This, in turn, leads to the requirement to discriminate, for example by targeting kinetic action carefully.

This is a fudge. Again, hardly anyone ever admits they are being unreasonable. 'Reasonableness' sits alongside motherhood and apple pie as something so tautologically good that it adds only token weight to an argument. Of course, we should be reasonable in all that we do. Nobody argues for unreasonableness. But how could Black be more precise?

We can exploit the chink in Black's approach by going back to the example of self-defence, and asking how many assailants it is 'reasonable' to kill to protect oneself or to save another innocent person. Black cites Germain Grisez, and his theologically inspired view that, because identical situations require identical recommendations, no further choices are needed once the path is chosen. So, if lethal self-defence is justified against an assailant, it will automatically be justified against the next, and the one after that, and so on. Following Grisez, Black would say it is justified to protect yourself with lethal self-defence against a thousand assailants who came at you one at a time. So why would it be unreasonable to kill the whole thousand at once, if that's the least you can kill to survive, or to save the single innocent person you're protecting?

This is an example of consistency being at odds with proportionality. Is being reasonable about being consistent, or is it about being proportionate – or even somewhere between the two? There can be an answer to this; road designers and certain medical professionals have to make choices which involve rationing chances of survival every day. Judges who issue prison sentences can make similarly serious decisions, and with an extra requirement common to the battlefield: when comparing the demands of society versus the rights of a convicted felon, the judge must factor in different moral significance for the individuals involved. Criminals count for less, just like enemies viewed through a gunsight count for less. And just as the criminal law has to go beyond being reasonable, to adjudicate between fairness, mercy, proportionality, consistency, and other factors, Black's 'natural law' approach should, too. 'Be reasonable' is not specific enough to resolve many dilemmas faced in war. Black may respond that going beyond 'reasonable' can lead to an intricate calculus which is ill-suited to the soldier under fire, but an attempt should be made nonetheless.

Finally, while Black focusses – very commendably – on outcomes and effects, he does also cover duties and virtues. He describes virtues as a quality of character which matches our values. In new and poorly understood military situations, for which rules have yet to be developed, and outcomes are unknowable, character is all that is left to guide what soldiers do. Black is essentially right that the best way to judge these virtues is by the outcomes they seek.

Black extends his coverage of virtues to cover culture – the habits of a group, and their collective character. Correctly, he observes that culture is so often at fault when unethical acts are committed by people working for organizations. If the word count had allowed, Black's chapter would have been even better if he had extended his analysis to explain how culture should be improved; it is about more than accepting the sub-optimal outcomes of cooperation, which he uses to justify military obedience. Cooperation is easiest when people think alike, but this group-think also leads to bad outcomes. Militaries will be most effective when they can harness cognitive diversity, yet we should also want them to practice a certain ethical conformity, so all their members are all good. The 'Natural law' approach espoused by Black may yet solve the riddle of how to improve culture through its focus on intended consequences: both cognitive diversity and ethical conformity should lead to good outcomes. Black is on to something with his foray into culture; I hope his next excellent paper gives more detail on this.

For the soldier under fire in the trench, talk of military culture and reasonableness may seem academic. And there may be no time to ask, 'Is my intention good?' But, as this response has shown, they are also highly relevant, since all three may offer the soldier a route to safety as the munitions explode around him.

Conclusion

Academic discourse naturally focuses on flaws and failings. Hence, these three responses have concentrated on what each of my co-authors seems, to me, to have got wrong. But each of them got much right, too, including in their critiques of my own chapter. One insight which stands out particularly is relevant for my own quasi-utilitarian approach, and it is this:

Just as academic discussion can be unduly critical, military pronouncements can be overly positive. Officers of all ranks are prone to overestimating how well they perform and how much they understand the landscape in which they operate. Many imagine they know more about the likely outcome of different courses of action than they actually do. This is a documented phenomenon, common to many professions, and known as the 'overconfidence bias', but the effect can be particularly strong for service men and women. It can be magnified by battlefield emotions, factors around competition for promotion, and an enemy keen to mislead. Hence, personnel under fire may be some of the worst judges of whether they know enough to make a decision based solely on the likely outcomes. Military predictions of what can happen may be much less accurate than those who make them expect.

The most atrocious decisions in war happen when rules are set aside, and those who confidently set rules aside risk making atrocious decisions. The quasi-utilitarian approach in Chapter 4 did not factor in this cognitive bias. When compensation for it is made, it calls for practical checks and balances to minimize the occasions when rules are wrongly ignored.

Notes

1 The six common requirements for a 'just war' have two conditions based on outcomes – that war has a reasonable chance of success, and the requirement for proportionality; two based on duties – the requirement to exhaust peaceful avenues, and for a just cause; and two based on virtues – war must be initiated by a legitimate authority, and with the right intent. Decent arguments can be made that some of these conditions straddle two or even all three of outcomes, rules, and virtues.
2 For example, the just war requirement for a 'Legitimate Authority' is usually interpreted as a legal term. This means that when ethics and law diverge, the 'just war' theory knowingly departs from what is 'right'.
3 For more on this, see King (2008, 195–199).
4 This conclusion applies to states, too: once they believe they have established a reputation for sticking to a costly deterrence commitment, they are more likely to act expediently when they are challenged again. For more on this, see Mazarr (2018).
5 To be clear, this is not a repeat of the fallacy. It is the argument that some ethical drives are innate, inescapable, and triggered by a perception of facts – our moral revulsion at a horrific murder, for example. Those who uphold that moral sentiments are to be taken seriously, and that they don't emerge through something like evolution, should say how else they think moral values originate.
6 That is, those means which are inseparably needed to achieve the ends. Since the two automatically come together, it is artificial to consider essential means and ends as separate.

References

King, Iain. 2008. *How to Make Good Decisions and Be Right All the Time*. Bloomsbury.
Mazarr, Michael J. 2018. *Understanding Deterrence*. RAND.

9

BLACK RESPONSE

Rufus Black

What the different authors of this volume are all seeking is a better theory of military ethics.

There is a shared view that any theory needs to be able to integrate the ethical importance of principles to guide decisions, consequences, and character in different ways. I have added that it also needs to have a central place for intentions.

Equally, all the authors recognize that we need an approach to the use of these insights that is practical for military decision makers. Roger Herbert makes this case very effective when he unpacks the experience of the United States Navy in teaching ethics. He explains that in their experience however much students might enjoy a course where they learn about the different philosophical approaches to making ethical decisions it does not equip military personnel well for the actual task of making decisions in the line of duty.

My approach to this problem is to demonstrate how natural law theory has long provided a single coherent theory in which all of those elements of intention, principle, consequences, and character play an important role. Complementing this analysis, I provide a test as to what would constitute the ethics being practical for military application and demonstrate the way natural law theory fulfils that test.

The approaches articulated by Deane-Peter Baker and Herbert seek to coordinate the use of different theories that focus these concepts into a single decision-making process. Iain King goes further and seeks to find a conceptual resolution, which is a utilitarian and deontological approach in the form of a quasi-utilitarian model.

Central to what separates these approaches is a foundational disagreement between myself and Baker and Herbert as to what would constitute an answer to the question of what is a good approach to military ethics? Baker sets that out clearly at the end of his chapter. Herbert doesn't provide as explicit an account to

DOI: 10.4324/9781003312925-11

justify his broadly similar approach but implicitly seems to rely on a similar logic to Baker.

For Baker, an adequate solution is to provide 'a practical decision-making methodology'. He contrasts this with 'a theory of ethics'. He is clear that his approach is a 'methodology' not a 'theory of ethics'. While he doesn't provide an account of what he really means by this distinction, which is critical to his whole approach, he asserts that being just a 'methodology' relieves him of the need 'to reconcile the very different (and arguably, irreconcilable) philosophical ideas underpinning the deontological, consequentialist, and virtue-based approaches to ethics'.

In defence of not seeking a coherent philosophical underpinning for his approach, he argues that his Ethical Triangulation model 'is an attempt to make explicit a heuristic that reflects how applied ethicists tend, in practice, to engage with ethical challenges'. He claims, without providing any evidence, that this sort of approach 'is widespread in applied ethics (as opposed to academic philosophical ethics)'. Setting aside the accuracy or fairness of the sweeping rhetorical characterizations of 'academic philosophical ethics' and 'applied ethicists', if his observations were true there are applied ethicists who are prepared to sacrifice philosophical coherence for pragmatism that doesn't explain what makes that right or logically sound.

At a minimum, it is a basic rule of logic, which I assume even ethicists appealing to the 'applied ethicist' defence would agree with, that we should prefer logically consistent and coherent explanations to ones that are less consistent or coherent. If that is the case, we shouldn't be giving up on the quest for a coherent theory and settling for a 'practical' answer unless we can make the case that no such logical answer is possible. Baker and Herbert have certainly not made that case, nor called to their defence anyone who has with whom we could then contend. I would observe that even if you are going to argue for 'practical' as the appropriate requirement, you need to be very explicit, in a way which Baker is not, as to what the test for a theory being 'practical' is. In my chapter, I argue that a theory has to be logically sound, which I assume puts me in Baker's 'academic' category. However, I also argue that the theory needs to be practical and articulate a clear set of requirements as to what practical means in a military ethics setting.

It is intellectually problematic enough for anyone professionally committed to rigorous thinking not to seek a more philosophically coherent foundation for military ethics. What is more, I will argue that the conceptual problems created by the irreconcilable philosophical traditions underpinning the 'Triangulation' approach create sufficiently serious practical problems that even against the test of 'practical' and being a 'methodology' it is very problematic.

The first conceptual problem is with Baker's account of why a theory of military ethics needs to have both deontological and consequentialist elements to it. Baker's argument is founded and founders on reducing what he calls 'the social contract that defines the liberal state to just two fundamental drivers, including a deontological account of human rights and consequentialist concern 'to ensure the protection of those who fall within its [the State's] scope'. He considers that the 'virtuous tension' between these two forces 'is foundational to liberal democracy – to remove one or the other is to fundamentally destabilise and distort the nature of our political system'. This is a sweeping simplification of the complex intellectual history that lies behind the conceptions of liberal democracies that we find around the world. The notion that there is a single Western concept of a liberal democracy does no justice to the very real difference between the conceptions of liberalism held and embodied even in countries with strong intellectual ties like the United Kingdom, Germany, and the United States. If a more adequate treatment of this history admits even an additional concept or two into the foundations of liberal democracy, then his argument that deontology and consequentialism alone need to be the foundation of military ethics doesn't have a sound basis.

Even if deontology and consequentialism were the foundations for a liberal democracy, it doesn't follow as a matter of logical necessity that the military ethics of the liberal democratic state needs also to be founded on them. Baker seeks to establish that deductive argument by holding that 'the liberal state must unquestionably take strong cognisance of the human rights of individuals affected by war' and 'ensure the protection of its citizens' with the unstated implication that this can only be secured by deontology and consequentialism. What Baker doesn't establish, but needs to, is that only deontology and consequentialism can secure both the protection of rights and the State. Just war theory equally protects both outcomes and as I will show it doesn't belong to either of Baker's categories of theory.

As Baker's chapter progresses, we find a strange inconsistency with this basic argument. This is because as we will shortly see, if this tension between deontology and consequentialism is so foundational to a liberal democracy, why does Baker in practice ultimately resolve the tension by having the consequentialist account ultimately trumps the deontological account so that it offers no more than 'the rules' in a rules-based form of utilitarianism?

There is of course a more fundamental objection to Baker's logic, which is that ultimately what should determine the content of military ethics is sound reasoning itself, not the extrapolation of a political philosophy. One is left wondering in Baker's account how you would establish a sound basis for military ethics in countries that are not Western liberal democracies. Given that part of the project of military ethics is to secure more ethical conduct for war universally, then you would think a better foundation for military ethics would be an approach that is applied universally. Part of the enduring power of Hugo Grotius' foundational work *De Jure Belli Ac Pacis Libri Tres* on international law and just war is that, as I outline in my chapter, he sought foundations that transcended the nation state and their particular political philosophies.

The second related conceptual problem is the lack of any explanation for why 'teleological' traditions are excluded and why our 'intention' or our 'intended ends' aren't important, let alone central considerations in Baker or Herbert's approaches. The closest explanation we get is Hebert simply dismissively waving them away saying 'my seniors never graded me on my good intentions; they judged me on outcomes'.

Certainly, neither Baker nor Herbert makes any form of intention-centred ethical analysis a feature of their approach. If part of the test for 'practical' is whether or not it equips military personnel for the challenges they will actually face, it is a real problem not having made intention a centrepiece of their analysis. There are a wide range of practical decisions where the central test of the law and public opinion about whether the decision was 'right' is to ask what was 'intended'.

Paradigmatically, in any targeting decision or any choice about the amount of force to use, central to the questions asked if events have gone wrong is what was intended, 'did you intend to kill innocent civilians?', 'was your intention only to use proportionate force?' Actions will go badly wrong in conflict from time to time and the ethical test will not just be what happened but what was intended? While Hebert's military seniors may not have been interested in intentions, courts, especially those trying war crimes, typically are.

In this account of what is left out, it is noticeable that all three of the other authors give no attention to natural law theory other than a single reference in Herbert's chapter that it was a type of theory taught in the US Navy's initial programme. When you are claiming that your methodology is about taking 'cognise of the core ethical intuitions', to not take account of intuitions generated by the most long-standing form of moral analysis in the Western tradition, which has provided the most systematic treatment of war and which laid the foundations for just war theory, is surprising to say the least.

In what I assume was an attempt to avoid dealing with natural law theory, Baker seeks to distinguish between the just war theory and the just war tradition and he then asserts with no evidence cited that the just war tradition 'reflects no single tradition of ethics but is instead a pragmatic merger of a range of ethical considerations'. That is just factually wrong if the word 'tradition' is to have any meaning. If you read Question 40, Article 1 in Thomas Aquinas's *Summa Theologiae* (perhaps we should cite this here?), whose systematization of natural law theory is foundational in Western ethical thought, you will see that the basic structure and concepts of what we commonly cite today as constituting the just war tradition are already clearly laid out. Aquinas explicitly builds on the tradition prior to him from St Augustine, which all belong to the same philosophical tradition of thought.

If you follow that tradition through its key moments in the works of Francisco de Vitoria's *De Indis* and *De Indis Relectio Posterior, sive de iure belli* and Francisco Suarez's *Disputatio de bello* and then finally into Hugo Grotius *De Jure Belli Ac Pacis Libri Tres* you can see unmistakably that just war tradition reflects a single tradition of ethics, which is a natural law tradition. Those ideas have certainly been used by subsequent thinkers who are not natural law theorists but any modifications they have made have not altered its fundamental ideas or structure to the extent you could plausibly claim as a material historical accuracy that it was 'a pragmatic merger of a range of ethical considerations'.

I fear these observations about just war theory reflect a reliance on authors whose interpretation of the history of Western moral philosophy is deeply and demonstrably flawed. Baker quotes Marcus Schulzke as 'correctly' pointing out that there are 'three dominant traditions in western moral philosophy: Aristotelian virtue ethics, Kant's deontological moral theory, and utilitarianism'. As I set out in my chapter, there is nothing correct about this assertion. It entirely ignores natural law theory, which as I set out provides the foundations of just war theory and the common law, and it well pre-dates and has a much longer history than either Kantian or utilitarian theories. Secondly, as I demonstrate, the division of moral theories into separate categories of deontological and virtue theories, and should they have been troubled by completeness, teleological theories, fails to account for those theories, like natural law theory, which belong to all three categories.

The third conceptual problem is that with no overarching account of ethics, there is no basis to determine when you apply one theory rather than another. This problem is very apparent in both Baker's and Herbert's accounts.

The key point in Baker's account is when he writes, 'the decision maker might find that the potential consequences of the course of action suggested by his or her assessment of the deontological principles are serious enough, and certain enough, that those consequences outweigh the deontological principles'.

Similarly, Herbert writes, 'If analysis of [deontological] constraints yields the result that an act is morally prohibited, it can't be morally justified by good consequences unless those consequences are exceptionally good or bad'.

In both cases, Baker and Herbert are asserting there is a principle, which says in effect 'you should follow a principle up until the harmful consequences of doing so clearly outweigh the good consequences'. We can legitimately ask, what is the basis of this principle? It is being put forward as a key principle to guide 'practical decisions' but we are not provided with any basis for why this is a sound principle.

If they say 'well it is a practical basis for making a decision' that doesn't avoid the question, why is it a sound practical basis for making the decision? In the end, they need to explain the basis of the principles that guide them. If they don't think that this is practical, then they are not equipping the soldier for the day when they have used this principle to make a controversial decision and they are being asked by superior officers, a military tribunal, a court, or a journalist, 'explain to me why you think there is a point where principles no longer hold true?' Baker and Herbert need to provide service people with a compelling answer to that question. I don't see one in their chapters.

I also assume from what they have written that this principle always holds true. If it doesn't, Baker and Herbert would need to explain why it doesn't and how we would know when to apply it or not and what the basis of that decision would be. Assuming it does hold true– and they give no indication that there are exceptions – then Baker's theory is not a form of triangulation rather it and the 'the Roadmap' in Herbert's model are both some form of rule utilitarianism. This is occurring because the deontological rules only hold true when the good consequences of adhering to them clearly outweigh the bad consequences. Baker and Herbert need to be clear about whether or not they are rule utilitarians of some form and whether they are not, explain when and on what basis some principles hold true irrespective of the consequences.

Whether or not they are really some form of utilitarian overall, utilitarianism is a form of thinking that has a central part in their 'practical' approach and it also plays an important role in King's quasi-utilitarianism.

That brings me to the fourth conceptual problem, which is the central role played by utilitarianism and the practical problems that creates. Herbert recognizes both the conceptual and practical problems. He observes, '[w]hile utilitarianism is a simple theory, determining net happiness is a demanding enterprise'. He sets out that short- and long-term consequences need to be considered, along with who is affected and the probabilities of the different outcomes. He then observes, 'A common critique of utilitarianism is that answering these questions requires omniscience'. Hebert is absolutely right and that takes us to the nub of the conceptual and practical problem of utilitarianism playing such a central role in their approaches.

One of utilitarianism's great strengths is that everyone's happiness (or whatever it is we are measuring, 'the good', preferences, etc.) counts and it counts equally. It is not time-bound, after all why does suffering tomorrow, next week or next year matter any less than suffering today? We know that in war, the physical and mental suffering of veterans and victims can go on for many years and is no less morally consequential than the suffering of a wound on the battlefield. The conceptual problem is that the need to understand the often widely fanning multiple chains of causation and to predict the future far exceeds human capabilities. Practically, especially in pressured decisions, any calculation of consequences will of necessity be constrained by time and cognitive load. Those problems are real enough before you consider the emotional distortions that can influence such reasoning.

Baker does respond to the critique others have made of the triangulation approach generally, that it is not realistic to think people will engage in such relatively sophisticated and involved cognitive exercise in high emotion, high pressure situations. He does so by trying to side-step the problem saying that even though this is an issue, there are plenty of situations when there won't be such pressure and that his method might reduce cognitive load. Neither is the answer to the critique of Ordiway, which he cites, or the more fundamental problem of calculating an answer accurately at all in any situation. While there will no doubt be occasions when soldiers can follow Herbert's advice, quoted by Baker, that 'you often have more time than you think – stop the bleeding, smoke a cigarette, and think it through' there will be other times when you don't. It is these times when the utilitarian calculus *practically* can't be done in the way required.

King in his chapter offers what sounds like a simpler solution, which he claims is quasi-utilitarian. He derives what he calls the Help Principle, 'Help someone if your help is worth more to them than it is to you' and claims that with this single principle you can answer any ethical dilemma. It sounds simple but in his book with the bold title, 'How to make good decisions and be right all the time' where he sets out a fuller account of this principle he explains that this is actually just a 'short-hand' for a fuller version 'Help someone if the all-time direct benefit of your help is worth more to them than it is to you'.[1] He then explains the six steps that are needed to apply this principle which are as follows:

1 Identify the options, noting that deliberately doing nothing is as much of an option as deliberately doing something.
2 Find out who will be affected by each option and how much.
3 Include effects which have already happened.
4 Filter out person-to-person wants.
5 Apply the reciprocity rule to each person.
6 Add it all up, and choose the best option. The option which maximizes all-time benefits is the best, so do it.[2]

These steps, which make clear how King's principle works, reveal that his theory is in practice a very lightly modified form of utilitarianism. To King's credit, his rigour makes it clear that if you use a utilitarian-type approach, then the process to determine what is the best option is an involved one.

A reasonable requirement for a methodology to count as 'practical' is that you are able to complete the decision-making process in the normal range of situations in which it will need to be applied, which in war includes decisions under real time and emotional pressure. The utilitarian dimensions of any of these approaches fail that test.

Herbert offers what he sees as a way around this problem. In a direct assault on the critique of the omniscience requirement, he simply says it is 'unjustified' because '[a]lthough utilitarianism sets a high epistemic bar, it asks only that we do our best given the information and time available'. If only it were that easy. The problem is the theory doesn't just ask us to 'do our best'. Our practical inability to determine whether someone will suffer harm doesn't make that harm any less significant in a utilitarian calculation of the right outcome.

If we think about a practical case, a soldier caught on a street with crossfire coming from multiple houses. They must make a very rapid decision. The soldier might decide to throw a grenade into the closest house from which shots are ema- nating, so they can take shelter there. If they are successful and enter the house, they might find that they have also seriously wounded all the members of the large family that are sheltering there. If they had more time to look carefully, they could have seen those civilians in there but they didn't. When they face the family, they can't say that their suffering isn't morally significant because they didn't have time to consider it. When someone does an after-action assessment and discovers the other houses, which could also have been reached with only slightly more risk, had no civilians in them, then from the utilitarian perspective the soldier's decision was a wrong one. Just because the soldier didn't know all the facts neither changes the suffering they caused nor the fact that their choice did not cause the least net suffering even if it is risk adjusted for the slightly greater risk to their own life of throwing the grenade into a different marginally more distant house.

I am not arguing the soldier made the wrong decision. The approach I offer in my chapter would justify their decision and importantly also explain the moral significance of the family's tragic suffering.

What I think Herbert is understandably doing is drawing on the ethical insight that we can only be responsible for consequences that we can 'reasonably foresee'. In my chapter, I explain the origins of this insight in natural law theory origins and how it is used. However, it is not an insight that comes with utilitarianism and any attempt to modify utilitarianism to incorporate it needs to set out how it can be done without compromising the conceptual basis of utilitarianism itself.

In a potentially similar move whose implications aren't clearly spelt out, Baker draws on what I think we are meant to interpret as a form of parallel between 'The Ethical Triangulation method' when it 'asks the decision-maker to consider "Consequences" and weigh those in light of deontological considerations' and the just war tradition. He claims the requirements to consider proportionality, likelihood of success, and last resort require 'the decision-maker to weigh the foreseeable outcomes of waging war against the right to do so'. If in his summary the word 'weigh' is to have any similar meaning to the way in which it is used when utilitarians use it, then his statement is very misleading. I set out how those requirements operate in just war theory from a natural law theory perspective and it does not involve anything conceptually similar to the type of 'weighing' of different sets of consequences that utilitarianism uses. What Baker does not seem to have in clear sight is that ethical theories can consider the potential consequences of decisions without being utilitarian.

What this analysis of Baker's and Herbert's work reveals is not only the conceptual problems with their triangulation-type approach but also that utilitarianism is a central component of their models and that they operate in practice as a form of rule utilitarianism or quasi-utilitarianism. King's more analytically rigorous approach means that he is clear that his approach is a form of quasi-utilitarianism. That greater clarity also enables us to see the further problems with utilitarianism that compromise all three approaches.

We have already seen that the approaches offered by Baker, Herbert, and King do not get around the omniscience objection. King's work clearly raises four further problems with the sort of utilitarian approaches found in all three authors' work.

The first of these is the incommensurability problem or the attempt to compare the incomparable. The problem is present right at the heart of King's core quasi-utilitarian principle: 'Help someone if the all-time direct benefit of your help is worth more to them than it is to you'. In King's explanation, the key phrase 'worth more than' is operationalized by the sixth step in his logic, which having established the options for choice and the consequences of them, is to '[a]dd it all up, and choose the best option'. The option which maximises all-time benefits is the best, so do it.[3] The real work here is being done by the concept 'the option which maximises the all-time benefit'.

To understand how this principle works, we need to know what 'benefit' means. When King introduces the Help Principle, he explains that 'help', which is the shorthand for creating a direct benefit, 'means doing anything which is valuable. . . . And since seeking value is what life is about, this advice can apply to almost everything people do'.[4] Had King explained that help or benefit meant a single outcome, like 'the level of life satisfaction that people experienced' and that we had some reliable scale to compare people's life satisfaction and each person counted for one, then if we were able to make all those assessments we could make a comparison.

A comparison would be possible because we are evaluating two sets of the same qualities. That is why the original utilitarian logic sought to find something comparable like 'utility' or 'happiness' and later 'preferences'. Mill and those who grasp the logic of his scheme recognized this was an essential feature of its coherence. While there are legitimate questions about whether those qualities accurately capture the nature of the ethical choice, it was at least conceptually coherent.

The problem is that King recognizes that we end up having to weigh different qualities. He provides the example of a group of people having to choose between going to see a film or taking their friend to a medical appointment.[5] The very reason a choice arises is because there are two differently valuable ends on offer – the complex bundle of satisfactions including emotional, intellectual, and aesthetic associated with seeing a film or attending to a person's health.

To resolve the choice, King says that for the person wanting to go to the cinema they would recognize in terms of helping it 'is worth more' for the person to go to their doctor's appointment. What the use of this example, which seems intuitively right, hides, is what the basis of comparison is when we say 'is worth more'. He explains, 'what matters is who needs something the most'.[6] This is just a question begging because he doesn't explain what 'needs something' means or how we establish what the 'the most' is. What King is effectively avoiding is explaining how you can coherently rank to differently worthwhile ends based on some intrinsic qualities they have.

The incommensurability problem is present in King's chapter where he applies the Help Principle to various war scenarios. He considers the example:

Major Howard is ordered to take a bridge from the enemy. The bridge is of crucial importance for his country's war effort, and he is lucky enough to have reliable and accurate intelligence about the enemy presence in the area. How should he proceed?

King observes that if it was just a question of the loss of lives, then his approach would involve choosing the course of action 'which minimises the number of life-years lost by his unit'. As an aside, if that is the calculation, it is not remotely practical because Major Howard has to weigh by the age of his soldiers how many years of life they might lose against some notional life expectancy. This also has the odd feature that the 'right' decision might depend on the average age of the soldiers and that in forces of developing countries with lower life expectancy there would be a bias towards increased risk of the loss of life.

Happily, we do not have to contend with these problems because King recognizes the choice is between two incomparable qualities, the lives of his soldiers and those of the enemies and the 'high purpose' of the pursuit of a just war. As King recognizes, the pursuit of a just war can't be reduced to a calculus of life-years lost. Justice doesn't have a price in human lives. His theory does not give a conceptual basis for how we can compare and rank these incomparable qualities. Rather, we need a careful principled non-utilitarian judgement to determine when it is reasonable to risk human lives in the pursuit of justice, which is exactly what just war theory and natural law theory do.

The second issue is that the use of a single ethical concept like the Help Principle or some other formulation of the utilitarian calculus risks dissolving important and hard-won ethical distinctions in the name of consistency with a single ethical idea. King provides an illustration of this problem when he argues in analysing a terrorism scenario that

> Few would consider the terrorists she is trying to beat a 'legitimate authority', which implies that countering terrorism requires a different approach to warfighting. However, quasi-utilitarianism suggests this distinction is artificial. Like the distinction between war and peace, the medieval notion that only some actors may use violence to seek a political effect is outdated.

King does not explain what is outdated about this notion that violence should only be used as a last resort by those with legitimate authority in defence of civil order and justice. Those who use force without legitimate authority therefore do two wrongs: the harm they do directly and the indirect harm they do by undermining the very authority needed to pursue the common good. However, for a quasi-utilitarian, notions like 'last resort' and 'legitimate authority' are at best summary rules that should be followed because generally that is what leads to a better outcome. The natural law theorist and others with principles-based theories would argue that these ideas are inherently important ethical notions that constrain and shape the use of force, not mere short-hands. Further, when we maintain them, we can more accurately describe the nature of the wrongs that occur when force is misused.

The attempt to collapse war and terrorism betrays another weakness, which is the risk which comes when we lose sight of the role of intentions in ethical theory. The reason terrorism is categorically different and should never be collapsed with war as a category is that war is capable of being conducted justly whereas terrorism is not. It is of the essence of terrorism that it seeks to achieve its aim by terrorising innocent people. Acts of war could be acts of terrorism, but they can also be just acts. Acts of terrorism will always be unjust. If you are a natural law theorist, they will also always be wrong. It is not a surprise that anyone with a utilitarian component in their thinking would not seek to keep that protection of the intentional killing of the innocent absolute because they can always think of cases where it can be justified.

This brings us to the third issue with utilitarian theories: they can be used to justify killing the innocent and using people as a means to an end. King to his credit does not avoid this implication. He poses the scenario:

Sven, a medic captured in the Bosnian war, is given a revolver and instructed to kill one of his fellow prisoners-of-war. If he does not, he is told ten POWs will be killed. Should he pull the trigger?

King's answer is clear:

What would Sven do, if he takes quasi-utilitarian advice? If the scenario is exactly as stated, then Sven should apply the rule about rules: he should kill one POW to save ten, because breaking his commitment to not kill, even when he is discovered, and even when factoring in his self-loathing for the deed and the undoing of his virtue, brings about a much better outcome than allowing ten POWs to die.

We should be clear here that King is arguing not just that in this case it is ethical to kill an innocent person, but he is making it very clear that when you build utilitarian principles inside your ethical approach as he, Baker, and Herbert all do, then there are times when it is the 'right' action to kill innocent people. We saw earlier that Herbert explicitly acknowledged there were times when exceptionally good or bad circumstances could lead to breaking what otherwise seem like solid principles, revealing of course they only appeared solid and ultimately presented no firebreak to the 'ethical' killing of innocent people.

One of the tests for a theory being practical, which I set out in my chapter, was the requirement that military ethics, '[a]lign the behaviour of service personnel with the legal constructs that govern the conduct of armed conflict'. Unless we do that, then we risk putting military personnel in the situation where we are saying to them, about enormously consequential life and death decisions, that what you did was ethical but it was illegal to the point of being a war crime. If we are going to seriously propose that kind of ethical theory, then those doing so should also as a matter of logical consistency be committed to arguing that the laws of armed conflict need to be reformed to better reflect ethics. I don't see Baker, Hebert, or King doing that. Rather, they risk placing military personnel who follow their 'practical' approaches in an invidious situation.

It is another of these practical requirements that I propose in my chapter that gives rise to the last of the objections to these types of utilitarian or quasi-utilitarian theories. I propose that it is an important practical requirement to:

Provide an account of authority that explains when it is ethical to follow orders, including the basic order to go to war, even when there are doubts about whether the order is ethical and when it is right to disobey those orders.

King underscores the importance of this requirement with his last practical scenario:

> · Captain Ivan is instructed to fire his artillery twenty miles west, to a location he thinks may be a school, and he has limited trust in his seniors. Should he fire as instructed?

King's analysis of the answer to this question is very problematic. The authority of the instruction is the central question. King does not provide a clear explanation of why the order of a superior has authority in the first place but he does maintain that it is conditional. He names two conditions.

The first condition is that the orders don't direct someone to break international law and the second is that the person receiving the order is paid. The requirement to be paid is odd because it suggests that authority is in some way contractual. Whereas what authority seeks to explain is why should someone do something? Is it because they are directed to or because it is the law irrespective of whether they agree with it?

Regarding the first condition, given that in the previous scenario King is able to envisage situations where breaking international law is in the greater good, it is not clear what the basis of this absolute statement is. It is inconsistent with the rest of King's theory.

In what seems like a contradiction to this statement, in the next paragraph King goes on to argue that '[f]rom time to time, people have to do bad things: in those cases when the behaviour of others means any better actions would lead to a worse outcome'. We seem to be back in the utilitarian mode where you could obey an order to break international law if that led to what you believed would be the best outcome. King then argues that '[o]nly when something greater is at stake should we jettison our personal integrity'. If in King's scheme the person is taking the action that leads to the best overall outcome, then what they are doing is the right approach and they are not jettisoning their personal integrity.

What these sorts of slips betray is an intuition that even when something is for the greater good, people can reasonably think both that it is the wrong decision and that they should go along with it. King's quasi-utilitarianism doesn't support that possibility. Even if it did, King's solution leaves people in the invidious and contradictory situation of having to ethically agree to do what is wrong. That is the sort of ethical framework that creates a significantly elevated risk of moral injury. I argue that one of the practical requirements of military ethics should be to reduce not increase the risk of moral injury. To solve these sorts of authority issues and to reduce the risk of moral injury require a very clear account of how authority works. I provide an outline of that account in my chapter.

King concludes his analysis by saying '[u]ltimately, Captain Ivan fires his artillery to make the world less hospitable to moral dilemmas'. Even on King's analysis that doesn't seem to be the ultimate reason why Captain Ivan fires the shell and if it was then it would be an unsound reason. The reason Captain Ivan fires the shell is because as he understands it, he received an order that has authority. If it was because he wanted 'to make the world less hospitable to moral dilemmas' then we are seriously misdescribing the ethical reality in front of us.

I admire the commitment of Baker, Herbert, and King to produce a practical form of ethics for military personnel. It is an important project. I have argued here that in their current form, their solutions are neither practical nor conceptually sound. A way forward would be for us to agree about what counts as practical. I have set out what a possible test for that could be and would look forward to seeing a good discussion refine it. On the conceptual front, we all agree on many of the core types of ethical insight that are needed. Rather than giving up on the task of conceptual coherence, I would invite Baker and Herbert to join King and myself in the quest to solve that challenge. For the sake of our military personnel and a commitment to the just use of force, I think we have an obligation to keep trying.

Notes

1 Iain King, *How to Make Good Decisions and be Right All the Time: Solving the Riddle of Right and Wrong* (Bloomsbury, 2008), 169.
2 King, *How to Make Good Decisions*, 169–170.
3 King, *How to Make Good Decisions*, 169–170.
4 King, *How to Make Good Decisions*, 109.
5 King, *How to Make Good Decisions*, 159–160.
6 King, *How to Make Good Decisions*, 160.

References

King, Iain. 2008. *How to Make Good Decisions and Be Right All the Time: Solving the Riddle of Right and Wrong*. Bloomsbury Publishing.

AFTERWORD

A Concluding Reflection on Military Ethical Decision-Making

David Whetham

I count it as a genuine privilege to be invited to offer a concluding reflection for this important volume. There is no doubting the critical importance of effective decision-making for military personnel. As this volume is being written, the Ukrainian Armed Forces are currently in possession of the moral as well as the military high ground in their struggle to defend their territory and people from Russian invasion. However, there is little doubt that their critical vulnerability is not the heavy armour or fast jets that much of the media focuses upon, but rather the ability to maintain international public support. At the moment, while there are many more states than we might have expected taking a cautious, neutral stance, there are also, fortunately, many who enthusiastically support Ukrainian efforts because they, and their domestic populations, are of the view that the Ukrainians are clearly the side in the right. just war thinking is predicated on a moral division of labour that, traditionally at least, separates the reason for fighting a war (*jus ad bellum*) from the way it is conducted (*jus in bello*), arguing there is a logical separation between the two levels of war. Whatever the arguments on either side as to the justice of their position, both sides are still obliged to conduct themselves according to the rules *in bello*. That does not mean that the two levels are completely independent, however. One cannot make an unjust war just by fighting it well, but one can certainly undermine a just war by fighting it badly (Walzer 1977, 210). This is the risk that Ukraine faces right now. If it were to ignore the rules of war, flout international conventions, and dismiss the importance of professional military ethics, it could very quickly find that international support drying up – why should people support one side over another if they are equally as bad? The Clausewitzian Trinity reminds us how public support is linked to the ability to conduct and sustain military endeavours, and the Ukrainian case study demonstrates in a very real sense the potential costs at the international level of getting military ethics wrong.

DOI: 10.4324/9781003312925-12

Law and ethics are, or at least should be understood as, inextricably linked. Proficiency in International Humanitarian Law (IHL) is a crucial requirement for members of all armed forces. However, such proficiency does not necessarily imply the ability to translate the legal standards into practice on the battlefield. While the law provides the parameters for legitimate action, it does not on its own actually suggest a course of action within those parameters – the law tells you what you are permitted to do, while an understanding of ethics helps answer the question 'What *should* I do?' This is why exploring the normative context is so important – applied military ethics is a medium that helps prepare military personnel to face possible ethical challenges and dilemmas in the most appropriate way. But which approach to ethical decision-making is best?

I want to stress from the outset that I have no intention of serving as the adjudicator for this debate, and nor have I been asked to. I leave that to the reader! Instead, my purpose here is, first, to briefly summarize the main points of agreement and the main points of disagreement between the four contributors to the debate. I then point the way forward on what I think are key areas of future research and debate, before finally adding some thoughts of my own on military ethical decision-making.

While this book is a debate between competing viewpoints, it is heartening to see how much agreement there is between the different contributors, which itself speaks well for the state of military ethics scholarship. Notably, the contributors are unanimous – and I wholeheartedly agree – on the need for a military ethical decision-making model that is 'credible enough to be trusted; comprehensive enough to cover the full range of dilemmas; and coherent, so that it gives clear rather than contradictory advice' (King) and which must, furthermore, be suitable for use in situations of high stress and under time pressure. All agree, furthermore, that the three mainstream approaches to ethics that dominate the Western landscape – deontology, consequentialism, and virtue ethics – each suffer from limitations that make them individually, or at least independently, unsuited to the task. King succinctly summarizes this shared view by noting that 'A virtue theorist looks only to the character of the agent; a Kantian studies only their actions; while the utilitarian looks solely to the outcome they create'. But each of the contributors is at the same time quick to point to the importance of military ethical decision-making taking into account all of these core ethical considerations. This kind of Rawlsian reflective equilibrium (or what non-philosophers might call common sense) is a core part – explicitly or often implicitly – of the way military ethics is taught in many military institutions around the world. Often it is not labelled as military ethics at all, falling under broader engagement with leadership or other military core proficiencies, but that does not mean it isn't there.

The contributors agree, also, that whatever model of ethical decision-making is adopted, it must be compatible with the just war framework (or as Black puts it, the 'underlying ethical logic that governs the laws of armed conflict and international law more broadly'), though as mentioned they are united in the view that the approach adopted must at the same time be capable of addressing a much wider range of ethically challenging situations than the combat-focused principles of the *jus in bello* and other relevant parts of the just war tradition can comfortably be applied to.

Where the contributors centrally disagree is as to whether what is required is a particular *theory*, one which provides 'philosophical rigour' (Black) or whether a decision-making framework that draws on multiple theories is a better option. Rufus Black and Iain King both opt for the former, while Deane-Peter Baker and Roger Herbert stump for the latter. There is an important question of political philosophy underpinning the division here. Is it, as Baker avers, somehow *inappropriate* for the military of a liberal democratic state to adopt a particular ethical theory as the basis of ethical decision making for its personnel? Or is it instead the case that military forces should be driven to identify and adopt 'a single theory that can coherently integrate ethical concepts of ends, principles, duties, virtues and consequences' (King)?

Aside from the specific details of the approaches they put forward, there are other differences in emphasis that are worth noting. Herbert's model gives greater explicit weight to elements like 'moral perception' and time pressure, though all the contributors certainly agree with the importance of these factors. Black stands out in the emphasis he gives to the need for the military ethical decision-making model or theory to

> provide an account of authority that explains when it is ethical to follow orders, including the basic order to go to war, even when there are doubts about whether the order is ethical and when it is right to disobey those orders.

Baker's approach is distinctive for the secondary role he gives to the place of virtue/character, where the others see this as of primary importance. King is unique among the contributors for offering an ethical theory of his own design.

It is difficult to assess, without further exploration, whether the deliverances of each of these approaches would, in general, be similar or whether applying them might lead to significantly different decisions in like circumstances. One data point in this regard is the tension between the position taken in the 'Moral Deliberation Roadmap' designed by Herbert and his colleagues that, 'except in extraordinary situations, general obligations trump special obligations', on the one hand, and King's model in which, 'since we experience empathy with individuals, not with an aggregate – that is, "one-to-one empathy rather than empathy for a group or for the greater good" – we are obliged to maximize happiness within a pair', on the other. This tension suggests that, in some circumstances at least, the 'Moral Deliberation Roadmap' and the 'Help Principle' might point the decision-maker in different directions.

This leads to my thoughts on where this volume leads us regarding future areas of research and debate. I agree wholeheartedly with Baker – as I'm sure do the other contributors – on the need for empirical work on the comparative practical effectiveness of these and other competing models of military ethics decision-making. Getting this right literally has the potential to save lives and at the same time mitigate the potential harm of moral injury among those who serve in the military. We owe them no less than the best decision-making tools we can conceive of. Deciding which theories, models, and approaches should be taken into consideration in this regard does require first satisfactorily answering the question Baker's contribution to this project raises, namely which of them are, and are not, appropriate for the military arm of a liberal democratic state.

This important discussion is not one that should be settled by philosophers and academics – if military service means that one is part of a genuine profession rather than simply an armed bureaucracy, critical reflection on its own values and standards is an essential part of that existence. While a healthy discussion should draw on both internal and external expertise, the discussion should be driven by, and most importantly, owned by, the profession itself.

I argued in the lessons learnt section of the Afghanistan Inquiry, that it is imperative that routine ethical discussion is normalized within military ethos. Routine critical reflection on the values and standards of one's organization – themselves institutional expressions of those virtues that are deemed to be of relevance – represent part of the process through which one acquires and practices *phronesis*. That is, the practical wisdom to determine what the right thing to do is in a particular situation or context. An appropriate, agreed, shared, and embraced decision-making tool can greatly assist in contributing to that process, which itself should be an ongoing one, not one that is simply engaged with as a new recruit or junior officer. For example, it is easy to agree that courage is a core military virtue, but even though courage is a value or a virtue that is supposedly easy to understand,

> what courage looks like on a patrol in Helmand or Uruzgan Province may be very different to the courage required by an administrator who wants to question the receipts submitted by a commanding officer, or the chief of the defence force when faced with a questionable direction from the Prime Minister.
>
> *(Afghanistan Inquiry 2020, 528)*

The difficulty lies not in knowing when to be courageous, but in determining what courageous actually looks like across the variety of different ethical challenges posed by contemporary military service.

The different ethical decision-making processes set out so clearly in this volume are an excellent basis for both prompting and grounding, those internal discussions, and debates.

References

Inspector-General of the Australian Defence Force. 2020. *Afghanistan Inquiry Report*. Department of Defence.

Walzer, Michael. 1977. *Just and Unjust Wars: A Moral Argument with Historical Illustrations*. Basic Books.

INDEX